Principles of Oceanography

Principles of Oceanography

Christopher Howard

www.statesacademicpress.com

Published by States Academic Press,
109 South 5th Street,
Brooklyn, NY 11249, USA
www.statesacademicpress.com

Principles of Oceanography
Christopher Howard

International Standard Book Number: 978-1-63989-439-0 (Hardback)

This book contains information obtained from authentic and highly regarded sources. All chapters are published with permission under the Creative Commons Attribution Share Alike License or equivalent. A wide variety of references are listed. Permissions and sources are indicated; for detailed attributions, please refer to the permissions page. Reasonable efforts have been made to publish reliable data and information, but the authors, editors and publisher cannot assume any responsibility for the validity of all materials or the consequences of their use.

Trademark Notice: Registered trademark of products or corporate names are used only for explanation and identification without intent to infringe.

Cataloging-in-Publication Data

Principles of oceanography / Christopher Howard.
    p. cm.
Includes bibliographical references and index.
ISBN 978-1-63989-439-0
1. Oceanography. 2. Ocean. 3. Earth sciences. 4. Marine sciences.
I. Howard, Christopher.
GC11.2 .P75 2022
551.46--dc23

TABLE OF CONTENTS

Permissions

Index

PREFACE

The study of various physical, chemical and biological features of oceans is known as oceanography. It is an inter-disciplinary field of physics, geology and biology which attempts to study the history of the oceans, their current situation and future implications of climate change. The subject of oceanography can be further divided into physical oceanography, biological oceanography, chemical oceanography and geological oceanography. Physical oceanography focuses on the properties and movements of sea water. Biological oceanography studies the biology and ecology of marine organisms such as plankton, bacteria, fish, etc. Chemical oceanography deals with the composition of seawater and biogeochemical factors that affect it. Geological oceanography is concerned with the geology of ocean floor and basins. This book is compiled in such a manner, that it will provide in-depth knowledge about the theory and practice of oceanography. Some of the diverse topics covered herein address the varied branches that fall under this category. Those in search of information to further their knowledge will be greatly assisted by this book.

A short introduction to every chapter is written below to provide an overview of the content of the book:

Chapter 1 - The branch of science which deals with the physical and biological study of oceans is known as oceanography. There are various branches of oceanography such as physical oceanography, chemical oceanography, biological oceanography, geological oceanography, etc. This is an introductory chapter which will briefly introduce all these significant branches of oceanography; **Chapter 2** - Biological oceanography refers to the study of ocean ecology which is focused on the interaction of organisms with various aspects of the oceanographic system. It includes the study of marine ecosystem, and marine flora and fauna. This chapter closely examines the key concepts of biological oceanography to provide an extensive understanding of the subject; **Chapter 3** - The scientific study of motions, physical properties, conditions and processes within the ocean waters is termed as physical oceanography. Some of the topics studied under physical and chemical oceanography are sea water properties, pressure in ocean water, ocean chemistry, etc. This chapter has been carefully written to provide an easy understanding of the varied facets of physical and chemical oceanography; **Chapter 4** - Ocean refers to a continuous body of salt water that covers nearly three-fourth of the surface of the Earth. A few of the topics studied in relation to oceans are world's oceans and seas, ocean currents, ocean waves and ocean tides. The aim of this chapter is to explore the domain of oceans. These topics are crucial for a complete understanding of the subject; **Chapter 5** - The continuous and directed movement of ocean water produced by numerous factors that act upon water is known as ocean current. Some of the significant aspects associated with oceans are ocean acidification, ocean heat content, geophysical fluid dynamics, plate tectonics, ocean turbidity, etc. This chapter discusses in detail all these aspects of oceans; **Chapter 6** - Ocean observations involve monitoring various aspects of the ocean such as air temperature, precipitation, surface wind stress, water vapor, sea surface temperature, sea surface salinity, backscatter, turbidity, etc. This chapter closely examines ocean observations as well as the ocean reanalysis method to provide an extensive understanding of the subject; **Chapter 7** - Pollution in oceans can be caused by humans when industrial, agricultural or residential waste enters the ocean and causes harm to the marine life and ecosystem. It can also be caused by noise, excessive amount of carbon dioxide, or the entry of invasive organisms. This chapter closely examines the effect of pollution on oceans to provide an extensive understanding of the subject.

I extend my sincere thanks to the publisher for considering me worthy of this task. Finally, I thank my family for being a source of support and help.

Christopher Howard

Chapter 1

Fundamentals of Oceanography

The branch of science which deals with the physical and biological study of oceans is known as oceanography. There are various branches of oceanography such as physical oceanography, chemical oceanography, biological oceanography, geological oceanography, etc. This is an introductory chapter which will briefly introduce all these significant branches of oceanography.

Oceanography is the study of the physical, chemical, and biological features of the ocean, including the ocean's ancient history, its current condition, and its future. In a time when the ocean is threatened by climate change and pollution, coastlines are eroding, and entire species of marine life are at risk of extinction, the role of oceanographers may be more important now than it has ever been.

Indeed, one of the most critical branches of oceanography today is known as biological oceanography. It is the study of the ocean's plants and animals and their interactions with the marine environment. But oceanography is not just about study and research. It is also about using that information to help leaders make smart choices about policies that affect ocean health. Lessons learned through oceanography affect the ways humans use the sea for transportation, food, energy, water, and much more. For example, fishermen with the Northwest Atlantic Marine Alliance (NAMA) are working with oceanographers to better understand how pollutants are reducing fish populations and posing health risks to consumers of the fish.

Oceanographers from around the world are exploring a range of subjects as wide as the ocean itself. For example, teams of oceanographers are investigating how melting sea ice is changing the feeding and migration patterns of whales that populate the ocean's coldest regions. National Geographic Explorer Gabrielle Corradino, a North Carolina State University 2017 Global Change Fellow, is also interested in marine ecosystems, though in a much warmer environment. Corradino is studying how the changing ocean is affecting populations of microscopic phytoplankton and the fish that feed off of them. Her field work included five weeks in the Gulf of Mexico filtering seawater to capture phytoplankton and protozoa—the tiniest, but most important, parts of the sea's food chain.

Of course, oceanography covers more than the living organisms in the sea. A branch of oceanography called geological oceanography focuses on the formation of the seafloor and how it changes over time. Geological oceanographers are starting to use special GPS technology to map the seafloor and other underwater features. This research can provide critical information, such as seismic activity, that could lead to more accurate earthquake and tsunami prediction.

In addition to biological and geological oceanography, there are two other main branches of sea science. One is physical oceanography, the study of the relationships between the seafloor, the coastline, and the atmosphere. The other is chemical oceanography, the study of the chemical composition of seawater and how it is affected by weather, human activities, and other factors.

About 70 percent of Earth's surface is covered by water. Nearly 97 percent of that water is the saltwater swirling in the world's ocean. Given the size of the ocean and the rapid advancements in technology, there is seemingly no end to what can and will be uncovered in the science of ocean-ography.

Oceanographic Explorations

Maritime History

Maritime history is the study of human activity at sea. It covers a broad thematic e lement of his-tory that often uses a global approach, although national and regional histories remain predomi-nant. As an academic subject, it often crosses the boundaries of standard disciplines, focusing on understanding humankind's various relationships to the oceans, seas, and major waterways of the globe. The concepts of nautical history records and interprets the past events involving ships, ship-ping, navigation, and seafarers. Many parts of their observations and understandings are taken for scientific studies or subsequent explorations and navigations. Oceanographic explorations were started way back around 4000 BC. The details available on various visible forms and oral-cum written records of sailors and explorers made the subject to grow as a separate science.

Ancient Explorations around 4000 BC

It was during the period around 4500 BC, the Ocean Diving habit was started by the ancient peo-ple. This was the time when, coastal cultures like those seen in Greece and China begun with people diving into the sea to gather food and engage themselves in commerce. In this period only, the ancient Egyptians developed their first sailing vessels. These vessels were probably used for sailing in the eastern Mediterranean Sea and near the mouth of the Nile River. It was also record-ed around 1000 BC, the Greek poet Homer mentions about fishermen diving up to 30 meters by holding on to a heavy rock.

First Sea Routes around 600 BC

In search of tin and other natural resources, the ancient Phoenicians develop typical sea routes around the Mediterranean sea , into the Red Sea and into the Indian Ocean. They also reached England by sailing along the western European coast. Treasure Diving was continued throughout these periods. Diving for treasure created many wars between the Persians and the Greeks. Around 414 BC the Diving was used in warfare. The Greek historian Thucydides writes about diving used in warfare in his narration of the siege of Syracuse.

First Crude Diving Bell Design

It was around 360 BC, the Greek philosopher Aristotle mentions about the use of a sort of crude, air supply diving bell. Aristotle wrote that "one can allow divers to breathe by lowering a bronze tank into the water. Naturally the container is not filled with water but air, which constantly assists the submerged man". It was around 325 BC, the first use of a Diving Bell was observed. Alexander the Great made use of a crude diving bell to employ combat divers during the siege of Tyre. The diving bell contained colored glass so that the divers could see through it. The divers used the bell to clear debris from the harbor. Alexander himself made several dives with the device to check on the progress of the work.

The Voyage of Pytheas

It was around 325 BC, the Greek astronomer and geographer, Pytheas, sailed towards north from the Mediterranean. He reached the coast of England. He was the first person, on record, who described about the land of midnight sun and the north of the Arctic Circle. He also developed methods for using the Sun and the North Star to determine the location of latitudes.

Circumference of the Earth Discovered

Around 200 BC, the Greek astronomer, Eratosthenes, became the first person to determine the circumference of the Earth. It was a breakthrough in the history of geography. He used the angles of shadows and the distance between Alexandria and Syene to arrive at a value of 40,000 km. The actual circumference of the Earth was found to be 40,032 km. The world's all geographic, earth and atmospheric sciences got a turning point after these observations.

Ptolemy's Map of the World

In the year 150 BC, the Greek astronomer and geographer, Ptolemy, produced the first map of the ancient world. It was showing the distribution of continents of Europe, Asia, and Africa as well as the surrounding oceans. This early map was one of the first known maps developed to include the lines of latitude and longitude. Around the year 100 BC, Salvage diving operations were continued around the major shipping ports of the eastern Mediterranean. It was around 200 AD, the First Indicated Use of Goggles was found from the artworks on Peruvian pottery. They have shown the divers wearing goggles and holding the fish with them. Fishery was an important occupation of people from time immemorial.

Viking Expeditions and Voyage of Leif Erikson

It was around 900 AD, the Vikings begun to explore and colonize the Iceland, Greenland, and Newfoundland. They were among the first groups to use the North Star to determine their latitude. In the year 1002, Norse explorer Leif Erikson became the first European to land in the North America. His voyage took place almost 500 years before that of the Christopher Columbus. He called the new land as Vinland and established a Norse settlement in what is now seen in the northern tip of Newfoundland in Canada. This paved the way for oceanographic explorations for territorial ownership, at a later stage.

Chinese Exploration

Chinese exploration includes exploratory Chinese travels abroad, on land and by sea, from the 2nd century BC until the 15th century. Before the advent of the Chinese-invented mariner's compass in the 11th century, the seasonal monsoon winds controlled Chinese navigation. In the year 1405, the Chinese made seven voyages consisting of over 300 ships with a combined crew of nearly

37,000. These voyages were designed to extend the Chinese influence and impress their neighboring states. Their economic pressures back home made them to put an end to these expensive voyages of the Chinese.

Indian Ocean and Beyond

Chinese envoys sailed into the Indian Ocean in the late 2nd century BC. They reportedly reached Kanchipuram, known as Huangzhi to them. During the late 4th and early 5th centuries, Chinese pilgrims like Faxian, Zhiyan, and Tanwujie, began traveling by sea to India, bringing back Buddhist scriptures and sutras to China. By the 7th century, as many as 31 recorded Chinese monks including I Ching managed to reach India the same way. In 674, the private explorer, Daxi Hongtong, was among the first to end his journey at the southern tip of the Arabian Peninsula, after traveling through 36 countries west of the South China Sea. Chinese seafaring merchants and diplomats of the medieval Tang Dynasty (618—907) and Song Dynasty (960—1279) often sailed into the Indian Ocean after visiting ports in South East Asia. Chinese sailors would travel to Malaya, India, Sri Lanka, into the Persian Gulf.

Voyages of Columbus and Vasco da Gama

The Spanish explorer Christopher Columbus sets out on his historic voyage across the Atlantic Ocean in search of a passage to China and India, in the year 1492. Instead, he discovered the North and South America, which eventually led to the establishment of European colonization in these newly discovered continents. Similarly, in the year 1498, they Voyage of Vasco da Gama made some unique findings. The Portuguese explorer, Vasco da Gama, sailed his ships around the Cape of Good Hope, located on the southern tip of Africa. He became the first European to reach India by boat. The expedition later returned to Portugal with a huge valuable cargo of rare spices and valuables. This has opened a new dimension in oceanographic explorations.

First Circumnavigation of the World

In the year 1519, Ferdinand Magellan and his fleet departed from Portugal to start a daring voyage of discovery, in history. The fleet became the first to sail around the whole world. Magellan's expedition of 1519—1522 became the first expedition to sail from the Atlantic Ocean into the Pacific Ocean. The Magellanic Penguin was named after him, as he was the first European to note it. In March 1505, at the age of 25, Magellan enlisted in the fleet of 22 ships sent to host D. Francisco de Almeida as the first viceroy of Portuguese India. Although his name does not appear in the chronicles, it is known that he remained there eight years, in Goa, Cochin and Quilon. He participated in several battles, including the battle of Cannanore in 1506, where he was wounded. In 1509 he fought in the battle of Diu. But unfortunately, Magellan could not live to see the entire accomplishment. He died on the Island of Mactan in the Philippines, in 1521, from the poison arrows sent by the local natives.

First Plans for a Submarine

An English mathematician, named William Bourne, drew up the first known plans for an underwater boat, in the year 1578. These plans called for a leather-covered wooden frame craft that would be rowed from the inside. Later, in the year 1620, the First Submarine was made. Dutch physician, Cornelis Drebbel, built the world's first submarine. The boat was made of wood reinforced with iron and covered with leather. It was a 12 seater submarine in two rows inside. They rowed with oars that stick out the sides through tight fitting leather sleeves to keep the water out. Drebbel made several trips in his submarine in the Thames River, near London, at a depth of about 4 to 5 meters.

Voyage of Edmund Halley

In 1690, the first Air-replenished Diving Bell was made. The English astronomer, Edmund Halley, developed a diving bell in which the atmosphere in the bell can be replenished by sending weighted barrels of air down from the surface. In 1698, Edmund Halley made what was called as the first scientific voyage to study the variation of the magnetic compass. During his voyages, he also made very important contributions to the understanding of the trade winds.

Waterproof Suit and Diving Devices

In the year 1715, the First Waterproof Suit was made. Chevalier de Beauve, a guard in the French Navy, developed a waterproof suit with lead shoes. Air gets supplied from the surface by two leather tubes fastened to the helmet. In 1715, the First Enclosed Diving Device was made. It was Englishman John Lethbridge who developed a completely enclosed, one-man diving dress. The device was made from a reinforced, leather-covered barrel of air, equipped with a glass porthole for viewing, and two arm holes with watertight sleeves.

The First Voyage of Endeavour

It was in 1768, Lieutenant James Cook, left the port of Plymouth, England on a voyage to observe a transit of the planet of Venus across the Sun. During this and other two voyages, he could explore and map the Pacific Ocean. He was the first person to use a chronometer to accurately determine his longitude at sea surface. In the year 1785, the American patriot and inventor, Benjamin Franklin, writes a lengthy letter to a scientific colleague in France. Known as his Sundry Maritime Observations, the letter announces the discovery of the Gulf Stream and touches on a wide range of maritime subjects such as ship propulsion methods, hull design, and causes of disasters at sea.

The Nautilus

In the year 1800, Robert Fulton, the inventor of the steamboat, built an early submarine called The Nautilus. This cigar-shaped craft was made of wood over iron plates, and used a horizontal rudder to control the up-and-down movement of the submarine. This system is still in use even today. President Thomas Jefferson signs a law establishing the United States Coast Survey, in 1807.

The Voyage of the H.M.S. Beagle

In the year 1831, the English naturalist, Charles Darwin, departed England aboard in the H.M.S. Beagle. The goal of the expedition was to perform a survey of Patagonia and Tierra del Fuego. Darwin studied the plants and animals at each new stop. He discovered many unique species on the Galapagos Islands off the coast of Peru in South America. These discoveries led to his groundbreaking theory of evolution. In his book, The Origin of Species, Darwin suggested that the deep ocean may be a sanctuary for living fossils. In the year 1840, the first Modern Sounding was done. Sir James Clark Ross conducted the first open ocean deep-water sounding in 2425 fathoms in the South Atlantic Ocean at at Latitude 27 S Longitude 17 W. The sounding was made using the traditional method of lowering a hemp rope over the side of the ship.

Observations of Early 19th Century

In 1842, Charles Darwin publishes The Structure and Distribution of Coral Reefs. In the year 1843, British naturalist Edward Forbes stated about his belief that the life cannot exist below 300 fathoms (1800 feet) in the deep sea. This declaration made a 20-year debate about the possible existence of a lifeless zone in the ocean known as an azoic zone. In the year 1849, Coast Survey soundings in support of Gulf Stream investigations resulted in the discovery of the continental shelf break and the continental slope. In 1853, Louis F. de Pourtales of the U.S. Coast Survey found indications of life in depths over 1000 fathoms (6000 feet), inside the deep sea. In 1857, the first Deep Sea Canyon was discovered by James Alden, a commanding officer of the Coast Survey Steamer Active, in the center of the Monterey Bay, off the coast of California. It is, now, known as the Monterey Canyon. In 1860, the first Chart of the Gulf Stream was published by the U.S. Coast Survey. In 1861, the first U.S. Navy Submarine, Known as the Alligator, was designed and used.

Observations of the Late 19th Century

In 1872, the most popular Voyage of the H.M.S. Challenger happened. The H.M.S. Challenger sailed from Portsmouth, England and begun its four-year trip around the world. During the voyage, scientists tested the salinity, temperature and density of the seawater. Information was also collected about the ocean currents, sediment, and meteorology. The crew discovered underwater mountain chains and hundreds of species previously unknown. This research was eventually consolidated into a fifty-volume research report known as The Challenger Report.

Other Developments

In 1874, the Sigsbee Sounding Machine was invented. This new machine became the basic model for wireline sounding in the deep sea for about 50 years. In 1882, the first Oceanographic Research Vessel, Albatross, was sailed. In 1888, the first Modern Electric Submarine, Gymnote, was designed by the French Navy. This steel-hulled craft was powered by a 204-cell battery. The Marine Survey of the Pacific, made in 1899, by the Swiss Zoologist Alexander Agassiz, was yet another milestone. In 1912, the Scripps Institute of Oceanography became affiliated with the University of California. The Scripps is one of the world's leading marine research centers, today. In 1914, the first Acoustic Exploration of the Sea Floor was attempted by the Canadian inventor Reginald Fessenden.

The Titanic Episode

On the 15th April 1912, the White Star Liner, Titanic, sunk after striking witan iceberg in the North Atlantic Ocean. Over 1500 passengers lost their lives during one of the worst peacetime maritime disasters in the history of mankind. This tragedy led to a concerted effort to devise an acoustic means of discovering the objects in the underwater zone, well ahead of a moving vessel. In 1985,

Dr. Robert Ballard, with the help of a tiny robotic submarine named, Jason, discovered the wreck of the Titanic. The wreck was found in 3850 m of water column at about 500 km off the coast of Newfoundland in Canada. Titanic was found in two separate pieces.

Mapping the Ocean Floor

In 1925, the German vessel, Meteor, sailed around the Atlantic Ocean and took detailed measurements of the ocean floor, using modern echo sounding equipment. These voyages revealed new information about the shape and structure of the ocean floor. Subsequently, in 1935, the Researchers at the Coast and Geodetic Survey invented an automatic telemetering radio sonobuoy. This instrument was considered to be the first offshore moored telemetering instrument. In the year 1937, Geophysicist and oceanographer, Athelstan Spilhaus, invented the first bathythermograph. It was a good measuring device that can continuously record the ocean water temperature.

The World War II Research

In 1941, during World War II, the electronic navigation systems were developed for precision bombing. A few years later, the Coast and Geodetic Survey, conducted its first hydrographic survey using these systems. Research during the war led to many new tools for ocean exploration, including deep-ocean camera systems, early magnetometers, sidescan sonar instruments, and early technology for guiding Remotely Operated Vehicles (ROVs). This was a turning point in all modern oceanographic explorations. In 1943, the Aqua-Lung, the first modern scuba system was designed. In 1951, the British ship, Challenger II, bounced the sound waves off the ocean bottom and located the sea's deepest point. It was subsequently named as the Challenger Deep.

Discovery of Mid-atlantic Ridge

It was in 1953, the American geologist, Marie Tharp, studied the sounding profiles from the Atlantic Ocean and discovered a rift valley. Later studies revealed that it is a continuous rift valley extending over 40,000 nautical miles along the ocean floor. This discovery provided the most useful evidences for the newly formed theory of continental drift, known today as plate tectonics. Later in 1961, the Scripps Institution of Oceanography developed the Deep Tow System. This sonar system became the forerunner of all remotely-operated and unmanned oceanographic systems, which is used even today. In 1963, the first operational multibeam sounding system was installed on the USNS Compass Island.

Robotic Submersibles

Alvin, a new deep submergence vehicle, was constructed in 1964-by the Woods Hole Oceanographic Institute. It was the first U.S. deep diving submersible and the first deep-sea submersible capable of carrying passengers. In 1965, the first Underwater Robotic submarine, Halibut, which can lower miles of lengths of its cables bearing lights, cameras, and other gear to spy on enemy armaments and find the other materials that were once lost on the bottom of the sea.

Deep Sea Drilling Program

In 1968, the deep sea research vessel, Glomar Challenger, departed on a 15 year expedition known as the Deep Sea Drilling Program. The ship criss-crossed the entire Mid-Atlantic Ridge between Africa and South America and took a lot of core samples. The ages of the samples provided solid evidences for the theory of seafloor spreading, which late gave rise to the modern theory of plate tectonics. In 1969, the first Long Duration Submersible Expedition was made off the coast of Palm Beach, Florida. Hydrothermal Vents were discovered in 1977.

Deep-sea Submersible-alvin

In 1977, the deep-sea submersible, Alvin, was sent to explore a part of the mid-ocean ridge north of the Galápagos Islands known as the Galápagos Rift. Alvin was following in the tracks of an nmanned vehicle, towed by the research vessel Knorr that had detected unusually high bottom-water temperatures and had taken photographs of odd white objects among the underwater lava flows— tantalizing clues about curious, possibly biological features.

Argo Project

In 1990, a program known as, Argo, was started by deploying 3,000 robotic probes throughout the world's oceans to monitor the climate, weather, and sea surface height. It was named after the mythical ship from the story of Jason and the Argonauts. The last probes were successfully placed in 2007. In 1993, the undersea Laboratory, Aquarius, began its operation off the coast of Key Largo, Florida.

Seafloor Mapping from Space

The worldwide mapping of the sea floor from space was done in 1995. In 2010, the Census of Marine Life was completed. This 10 year project involved 2,700 scientists from 80 nations. The census revealed what, where, and how much lives and hides were there in the global oceans. The data is made available in the online directory that allows anyone to map the global addresses of species, today. In 2012, a Japanese expedition and film crew captured the first video of a giant live giant squid in its natural environment.

Modern Trends in Oceanography

Citizens, sailors, and scientists have observed the seas and oceans for several hundred years. First from the shore, then from ships and submersibles, and recently from satellites. While carrying out these explorations, scientists and engineers learned that they could sometimes leave some sophisticated instruments in the ocean, secured by wires, buoys, weights, and floats and allow them to systematically map the ocean floors. This method is known as the moored observatory. This approach has advanced the understanding of the oceans and their interaction with the Earth and the atmosphere.

Underwater Vehicles

In order to understand the ocean, scientists often find they have to get themselves or their instruments into very specific parts of it. Traditionally, researchers have used ships to photograph the depths. They used to drop floats and drifters into the currents, and to collect samples of water, rock, and marine life, for their analysis. In recent years, the spectrum of available observing tools has grown to include human-occupied submersibles, remote-controlled vehicles, autonomous, and towed robots.

Earth's First Artificial Satellite

In the year 1957, the Soviet Union successfully launched its satellite, Sputnik I. it was the world's first artificial satellite. About the size of a basketball, and weighing only 183 lbs, Sputnik took about 98 minutes to orbit the Earth. This was the Earth's first artificial satellite ushered in an age of exploration not only of space, but of the Earth's land masses and oceans as well. Satellites that detect and observe different characteristics and features of the Earth's atmosphere, lands, and oceans are often referred to as environmental satellites.

Polar Operational Environmental Satellites

POES (Polar Operational Environmental Satellites) maintain an orbital height of about 500 mi and take about 100 min to complete an orbit. Sea surface-temperature is one important type of data that these satellites provide. Knowing the temperature of ocean water is an important aspect in oceanography. Temperature changes influence the behavior of fish. It can cause the bleaching of corals, and affects the weather all along the coast. Satellite images of sea-surface temperature also show patterns of water circulation.

Mapping Ocean Surface Indicators

In addition to temperature, satellites also provide information about the color of the ocean. This allows scientists to detect the presence of algal blooms, river plumes, and other events. Satellite imagery is also being used to map features in the water, such as coral reefs. Sensors such as Landsat7 and IKONOS provide detailed information on local areas. Landsat-7 is part of the Landsat program, one of the longest existing environmental satellite programs. Satellites providing environmental imagery are used jointly with other organizations that receive data from various sensors. For example, marine animals, such as sea turtles and manatees, can be fitted with transmitters that relay information about their location. Satellites are becoming a standard tool for studying the oceans.

Background of the SeaWiFS Project

The purpose of the Sea-viewing Wide Field-of-view Sensor (SeaWiFS) Project is to provide quantitative data on global ocean bio-optical properties to the Earth science community. Subtle changes in ocean color signify various types and quantities of marine phytoplankton (microscopic marine plants), the knowledge of which has both scientific and practical applications. The SeaWiFS Project will develop and operate a research data system that will process, calibrate, validate, archive and distribute data received from an Earth-orbiting ocean color sensor.

Ocean Color Reflecting Marine Biota

The concentration of microscopic marine plants, called phytoplankton, can be derived from satellite observation and quantification of ocean color. This is due to the fact that the color in most of the world's oceans in the visible light region, (wavelengths of 400-700 nm) varies with the concentration of chlorophyll and other plant pigments present in the water, i.e., the more phyto-plankton present, the greater the concentration of plant pigments and the greener the water. Since an orbiting sensor can view every square kilometer of cloud-free ocean every 48 hours, satellite-acquired ocean color data constitute a valuable tool for determining the abundance of ocean biota on a global scale and can be used to assess the ocean's role in the global carbon cycle and the exchange of other critical elements and gases between the atmosphere and the ocean.

Geostationary Operational Environmental Satellites (GOES)

GOES satellites provide the kind of continuous monitoring necessary for intensive data analysis. Because GOES satellites stay above a fixed spot on the surface, they provide a constant vigil for the atmospheric "triggers" for severe weather conditions such as tornadoes, flash floods, hail storms, and hurricanes. When these conditions develop the GOES satellites are able to monitor storm development and track their movements. GOES satellite imagery is also used to estimate rainfall during the thunderstorms and hurricanes for flash flood warnings, as well as estimates snowfall accumulations and overall extent of snow cover. Such data help meteorologists issue winter storm warnings and spring snow melt advisories. Satellite sensors also detect ice fields and map the movements of sea and lake ice.

The Landsat Program

The Landsat Program provides the longest continuous space-based record of Earth's land in existence. Since 1972, Landsat satellites have collected measurements of Earth's continents and surrounding coastal regions that have enabled people to study forests, food production, water and land use, ecosystems, geology, and more. The long data record allows scientists to evaluate the dynamic changes caused by both natural processes and human pra ctices. The Landsat Program is jointly managed by the U.S. Geological Survey and NASA. Every day, Landsat satellites provide essential information for land managers and policy makers to support wise decisions about our resources and environment in the places we live and work.

Global Ocean Biomes

The large ocean ecosystems are called as biogeographic provinces. Biogeographic provinces provide useful categories for comparing and contrasting important ocean processes such as primary production, carbon flux, and species distribution and diversity. Climatological provinces have been identified using a priori expert knowledge. The global remote sensing data automatically produce the time and space resolved province distributions. Today, 3-d mapping of ocean water masses and biomes are being done in oceanographic analysis. Satellite oceanographic products provide an unequaled view of the global ocean surface allowing us the opportunity to map important biogeo-chemical processes in the ocean such as primary production and carbon export to the deep ocean.

Ocean Exploration and Human Health

At least 20,000 new biochemical substances from marine plants and animals have been identified during the past 30 years, many with unique properties useful in fighting disease. "Biodiscovery" researchers have had success in all types of ocean environments. A 1991 expedition by the Scripps

Institution of Oceanography's Paul Jensen and William Fenical resulted in the discovery of a new marine bacterium, Salinispora tropica, found in the shallow waters off the Bahamas. This bacterium produces compounds that are being developed as anticancer agents and antibiotics. It is related to the land-based Streptomyces genus, the source of more than half of our current suite of antibiotics.

"Metagenomics"

Researchers are now looking at the ocean through a new lens: the science of "Metagenomics." In the ocean, as on land, thousands of species of tiny microbes play a key role in nutrient cycles, including the carbon cycle, and in maintaining Earth's atmosphere, among other important functions. Metagenomics enables researchers to quickly sequence the DNA of all microbes in a given sample, in this case seawater, revealing how microbes function and how they may work in different environments. After mapping 64 million base pairs (units of DNA), thousands of new genes and also variations in the genetic composition of microbes at different depths were also discovered. Genomes from organisms in the deep ocean had many "jumping genes," or pieces of DNA that can move from one part of the genome to another.

Human-occupied Submersibles

Two other U.S. academic and research organizations operate human-occupied submersibles: the University of Hawaii has two Pisces submersibles capable of diving to 2,000 meters (1.2 miles), and Harbor Branch Oceanographic Institution in Florida has two Johnson Sea-Link submersibles that can dive to approximately 1,000 meters (0.6 miles). France and Russia operate HOVs that can dive to a depth of 6,000 meters (3.7 miles), and Japan's Shinkai 6500 is able to dive down 6,500 meters (4.0 miles). China is building an HOV that will be able to dive to 7,000 meters (4.3 miles).

Autonomous Benthic Explorer

The Autonomous Benthic Explorer, more commonly known as ABE, is the first underwater robotic vehicle of its kind. ABE was designed and built at the Woods Hole Oceanographic Institution (WHOI) in the mid 1990's. ABE weights approximately 1200 pounds and is a little over 2 meters long. ABE's top cruising speed is 2 knots. As an Autonomous Underwater Vehicle (AUV), ABE is a true robot, able to move on its own without a pilot or tether to either ship or submersible. This gives ABE the advantage of covering large areas of underwater terrain. ABE was invented to address scientists' frequent need to monitor underwater areas over long periods of time. ABE is designed to perform a predetermined set of maneuvers, take photographs, and collect data and samples within an area about the size of a city block. After accomplishing its mission, ABE "goes to sleep," conserving its power supply for months at a time, allowing for future missions without recharging its batteries.

Satellite Communications

Although ships have been crisscrossing the ocean for centuries, Earth observations from satellites provided the first truly global view of the ocean and its processes. Recent improvements in satellite communications on ships are fundamentally changing the nature of sea-going science. Many oceanographic ships now have Internet connections through a network known as HiSeasNet. With this network, shipboard scientists can work in real time with their land-based colleagues.

International Indian Ocean Expedition (IIOE)

The International Indian Ocean Expedition (IIOE) resulted from a cascade of effects. The International Geophysical Year of 1957-1958 had shown the value of coordinated multinational efforts in ocean science. This realization resulted in the International Council of Scientific Unions (now the International Council for Science) creating the Scientific Committee on Oceanic Research (SCOR) to continue to stimulate international cooperation in ocean sciences. The International Indian Ocean Expedition (IIOE) was one of the greatest international, interdisciplinary oceanographic research efforts. It involved forty-six research vessels (under fourteen different flags) that carried out an unprecedented number of hydrographic surveys (and repeat surveys) of the entire Indian Ocean basin from 1960 – 1965.

Emergence of Innovations

In the last 50 years, since the formation of IIOE, there have been two fundamental developments in ocean science. The first one is the emergence of new components of the ocean observing system - most notably remote sensing and Argo floats. The second one is the emergence of ocean modelling in all its facets - including short-term forecasting, seasonal predictions and climate projections. These developments have revolutionized our understanding of the global oceans. Ocean exploration continues to illuminate details about Earth processes.

Ocean Altimetry

Spaceborne radar altimeters are superb tools for mapping the ocean-surface topography, the hills and valleys of the sea surface. These instruments send a microwave pulse to the ocean's surface and determine the time taken by the waves to return. A microwave radiometer corrects any delay that may be caused by water vapor in the atmosphere. Combining these data with the precise location of the spacecraft makes it possible to determine the sea-surface height very accurately. The strength and shape of the returning signal also provides information on the wind speed and height of ocean waves. These data are used in ocean models to calculate the speed and direction of ocean currents and the amount and location of heat stored in the ocean, which, in turn, reveals the global climatic variations.

Concepts of Oceanography

Basic concepts in oceanography include major wind patterns that drive ocean currents, and the effects that the earth's rotation, positions of land masses, and temperature and salinity have on oceanic circulation and hence global distribution of radioactivity.

Global Processes

Global Wind Patterns and Ocean Currents

The wind systems that drive aerosols and atmospheric radioactivity around the globe eventually deposit a lot of those materials in the oceans or in rivers. The winds also are largely responsible for driving the surface circulation of the world ocean, and thus help redistribute materials over the ocean's surface. The major wind systems are the Trade Winds in equatorial latitudes and the Westerly Wind Systems that drive circulation in the north and south temperate and sub-polar regions. It is no surprise that major circulations of surface currents have basically the same patterns as the winds that drive them. Note that the Trade Wind System drives an Equatorial Current-Countercurrent system, for example. There is a North Equatorial Current running from east to west in every ocean: Indian, Pacific and Atlantic. There is a South Equatorial Current, just into the southern hemisphere, running in the same direction, from east to west. There is an Equatorial Countercurrent, running in the opposite direction that essentially splits the two.

At about 40 to 60°S and N, the Westerly Wind systems prevail. The southern hemisphere westerlies essentially blow completely around the globe almost unimpeded by land. The current system established by these winds (West Wind Drift) thus is driven completely around the globe, almost unobstructed by land. Comparable to this, in the northern hemisphere, the North Pacific and North Atlantic Currents move from west to east. In the Indian Ocean there is no comparable northern westerly current because of the Asian land mass. The ocean flows coupling the equatorial and westerly current systems are the Kuroshio and California Currents in the north Pacific, and the East Australian and Peru Currents in the south Pacific. In the north Atlantic, the Florida Current/ Gulf Stream is comparable to the Kuroshio in the Pacific, and the Canary Current is comparable to the California. In the south Atlantic, the Falkland/Brazil Current and the Benguela Current connect the equatorial and westerly currents.

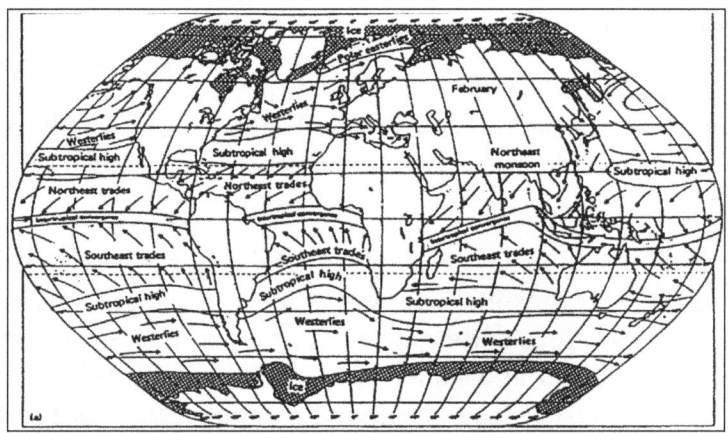

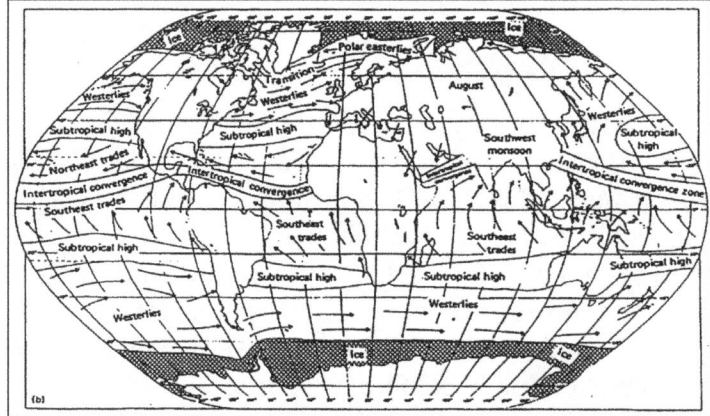

Figure: Prevailing winds over the ocean in (a) February, and (b) August.

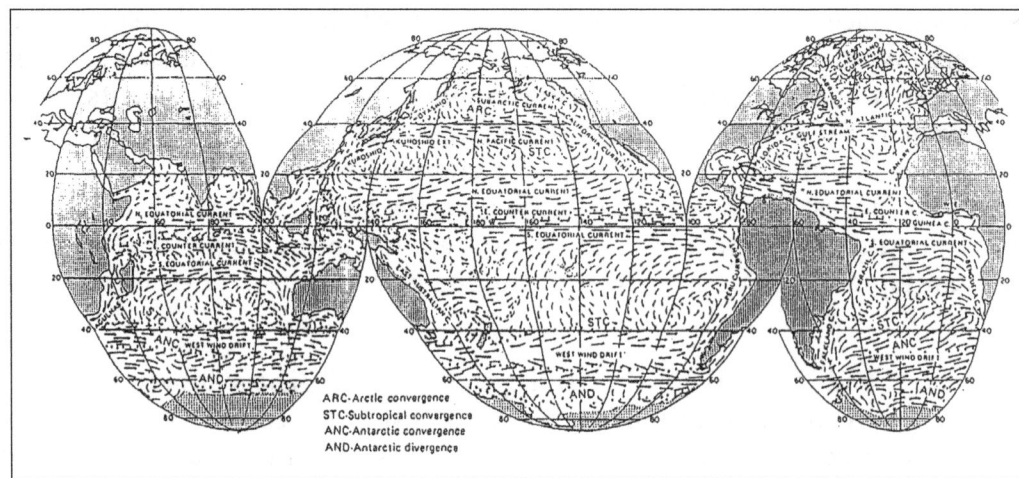

Figure: Surface ocean currents in February-March. Length and thickness of arrows denote relative current speeds. Note the major easterly and westerly currents, and the mid-ocean gyres.

In above Figure, one can see that the current systems circumscribe large oval regions in each ocean basin. These oval regions are the open-ocean gyres. The gyre systems essentially are the parts of the ocean usually described as oligotrophic, or with relatively little biological productivity in the surface waters (although recent work suggests they are more productive than originally thought). The gyres actually have a different topography than the regions of major currents. The sea surface is slightly higher in the center of each gyre than it is in other places, and that is important for distributions of properties, including radioactivity.

In the high polar regions, particularly in the northern hemisphere where there is no land mass comparable to the Antarctic continent, easterly winds (Polar Easterlies) move surface currents eastward; however, because of land interference, easterly currents are often turned into complex patterns.

The basic pattern is thus one of major wind systems driving surface currents around the perimeters of each ocean basin in each hemisphere on the earth's surface, piling water up a bit in the central gyres. Complexities in this simple pattern occur because of the positions of the continents with respect to the wind/current systems, and sometimes because of major changes in the wind systems themselves either on a seasonal basis (as in the Indian Ocean monsoons) or on less frequent time scales (as in El Nino-Southern Oscillation events).

Seasonal Monsoons

The most striking seasonal changes in current pattern usually occur in the Indian Ocean, and these are intimately tied in with monsoon winds. During summer (generally May to September), the land mass of Asia is greatly warmed relative to the adjacent Indian Ocean. When the land mass is warm and the ocean is cold, the continental air rises and draws the cooler air off the marine system. The air drawn from the ocean to the land creates the Southwest Monsoon. At this time of year surface currents in the western Indian Ocean move northward while those in the eastern Indian Ocean generally move southward to join the westward-moving South Equatorial Current. This Southwest Monsoonal circulation becomes very important from the standpoint of biological productivity. When the land mass of Asia cools in the winter, the situation reverses. The high heat capacity of the water causes the air mass above the Indian Ocean to warm, and thus to rise and draw the colder air off the land mass. This creates the Northeast Monsoonal circulation of the sea surface. The general pattern shown with the North Equatorial Current and Equatorial Countercurrent reformed, is thus re-established.

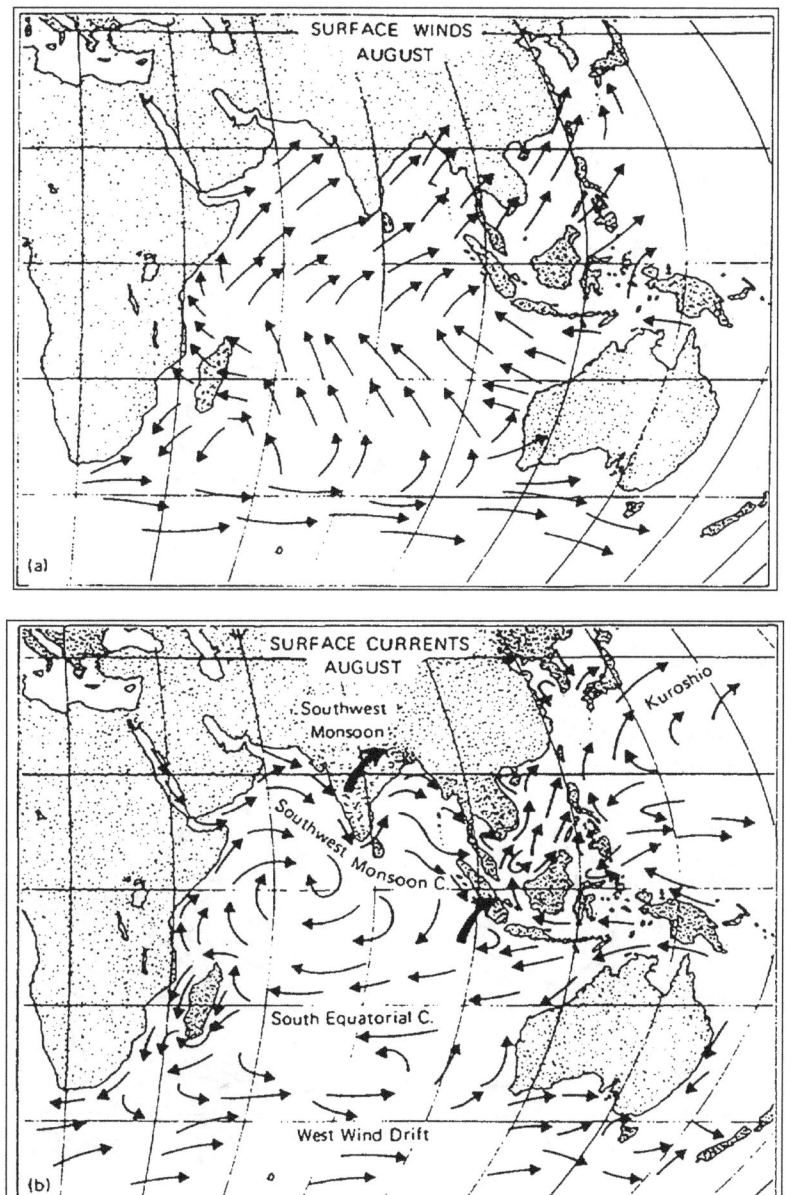

Figure: (a) Surface winds and (b) currents in summer during the Southwest Monsoon.

Other Seasonal Changes

There are other seasonal shifts that occur, which can be very important to the distributions and redistributions of radionuclides and other materials, and therefore to the potential effects these radionuclides may have on marine life. Patterns of sea ice change seasonally in both Polar Regions, for example. Seasonal changes in prevailing wind directions along coastlines other than those in the Indian Ocean often create movements of surface water away from or toward the coasts, thus either transporting materials away from the land or onto the beaches. However, it is important to recognize that meteorological high and low-pressure systems do vary in position by season over the surface of the Earth, thereby seasonally re-positioning major and minor wind systems, which in turn affect the positioning and strengths of most ocean currents on a seasonal basis.

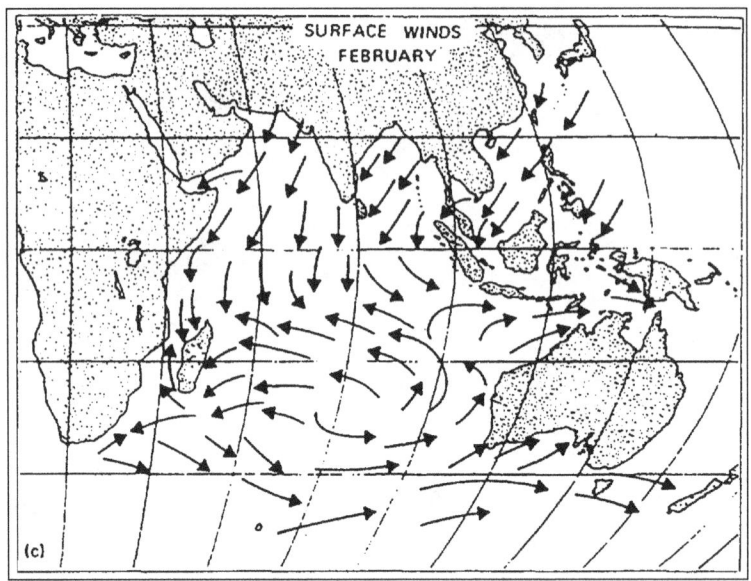

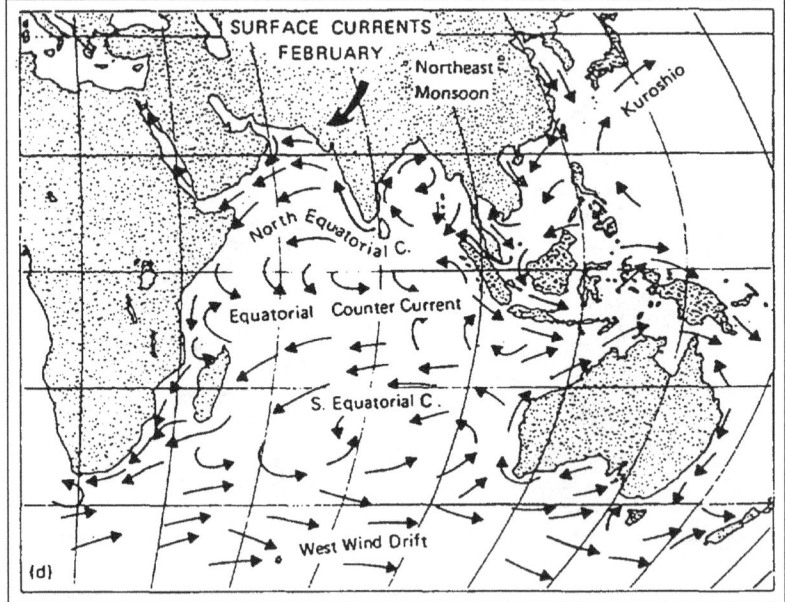

Figure: (a) Surface winds and (b) currents in winter during the Northeast Monsoon.

The Coriolis Effect

The Earth's rotation about its axis causes the winds and surface ocean currents to follow curved paths instead of the straight ones that we might witness if the Earth were not rotating. This apparent curvature is accounted for by the Coriolis Effect. To give an example of the Coriolis effect, let us presume that we have an air mass resting at the equator. While sitting at the equator, the air mass is moving east at 1670 km/hr, because that is the speed of the Earth's rotation at the equator. If the air mass is put into motion toward the North Pole (i.e., if we create a northward-blowing wind), that air mass still moves eastward at 1670 km/hr, in addition to its speed toward the North Pole. As the air mass proceeds northward it moves over portions of the Earth's surface rotating at ever decreasing speeds; e.g., at 30°N the Earth's surface is rotating at only 1446 km/hr, and at 60°N only 835 km/hr. The northward-moving air mass still maintains its 1670 km/hr eastward speed, however, and is therefore moving eastward at a speed faster than the Earth beneath it. To a person standing at the Equator, at the initial site of the air mass, the air mass appears to veer ever more to the right as it proceeds ever more northward. A person on a stationary platform in space, however, would see that the air mass actually travels in a straight line, with the Earth turning beneath it. Had the air mass at the equator been put in motion toward the South Pole, the deflection would still be eastward but would appear to the person standing at the equator to veer to the left, not the right. Had the equatorial air mass been put into motion along the equator rather than perpendicular to it, it would have moved in a straight line, with no veering. There is no Coriolis Effect at the equator.

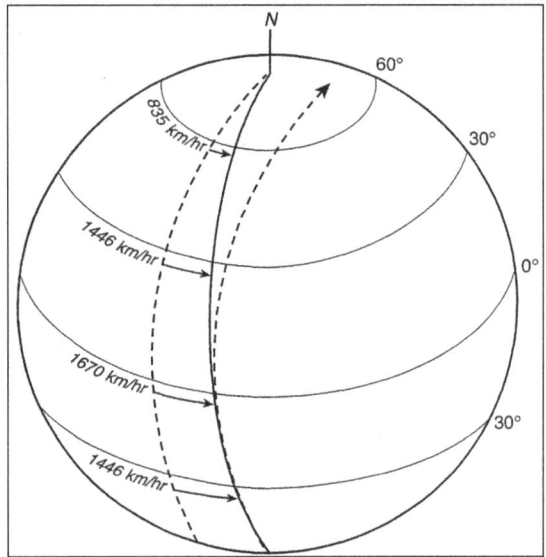

Figure shows apparent deflection to the right of a hypothetical parcel of water moving from the equator to the North Pole, due to Coriolis Effect. The deflection would be to the left for the water parcel moving from the equator to the South Pole. These deflections are due to the decreasing speeds of the Earth's surface at higher latitudes as the Earth rotates in an eastward direction.

The coupling of the winds to the surface ocean currents, given the Coriolis effect, makes the surface currents veer to the right of the wind direction in the northern hemisphere and to the left of the wind direction in the southern hemisphere. With the equatorial Trade Winds blowing from east to west, and the higher-latitude Westerlies blowing from west to east in both hemispheres, the global surface current systems tend to turn in a clockwise direction in the northern hemisphere and in a counterclockwise direction in the southern hemisphere. Seasonal changes in wind direction, such as the different monsoons in the Indian Ocean, can change the direction of surface flows.

Ekman Spiral and Ekman Transport

The ocean is a three-dimensional system, and distributions of surface-injected radionuclides are affected not only by atmospheric-oceanic interactions at the immediate surface, but also by the interaction of surface waters with water layers below. In the northern hemisphere, the average deflection of surface waters from the prevailing wind direction is about 45° to the right, due to the Coriolis effect (in the southern hemisphere it averages 45° to the left). Frictional drag of the surface layer on the next layer below it causes that next layer to decrease in current speed and to change direction slightly further to the right. Proceeding downward through succeeding water layers yields ever-decreasing current speeds with continual directional change to the right. At some depth the current, now greatly reduced in velocity, actually reverses direction, and eventually a depth is reached whereby no energy is left to produce any current at all. This spiralling phenomenon is called the Ekman spiral, after the scientist who first described it, and the depth range from the water surface to the depth of no measurable wind-induced energy is called the Ekman layer. Depth of the Ekman layer varies depending mainly on the wind strength imparted to the ocean surface.

If one calculates the net directional movement of all water in the Ekman layer, one finds that it is approximately 90° to the right of the surface wind direction in the northern hemisphere (90° to the left in the southern hemisphere). The amount of water transported in the Ekman layer over some horizontal distance has been called the Ekman transport. One can see how knowledge of Ekman transport would be vital to predicting distributions of surface-injected radionuclides throughout the top layers of ocean waters. In most cases the Ekman layer does not exceed about 100 m depths, so Ekman transport is still a near surface feature; however, the upper 100 m usually encompasses all, or most, of the lighted zone (euphotic zone) of the sea, where all photosynthetic production by single-celled algae (phytoplankton) takes place, and of course it is the interactive link between the atmosphere and the ocean, as we've seen. Thus, the Ekman layer is an extraordinarily significant, though small, part of the total volume of the global ocean, accounting for much of the dispersal of all manner of organic and inorganic materials.

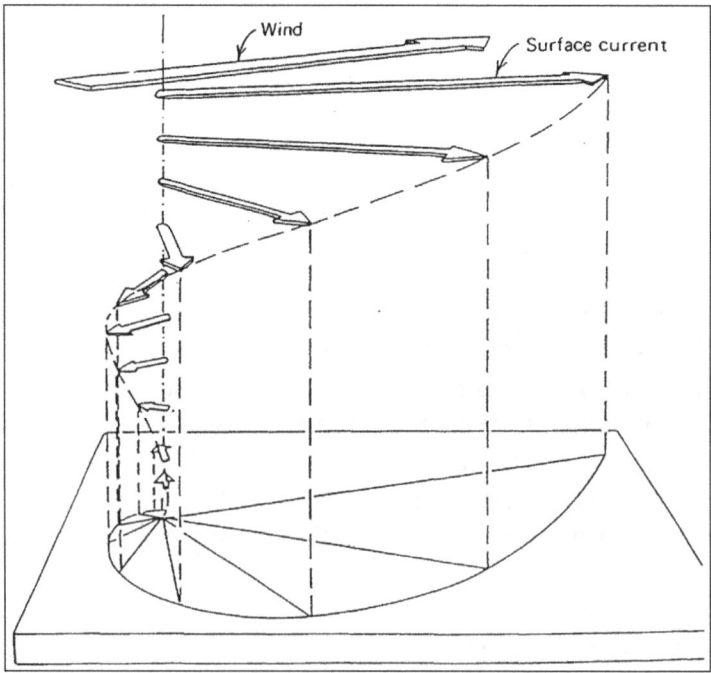

Figure: Schematic representation of the Ekman spiral formed by a wind-driven current in deep water.

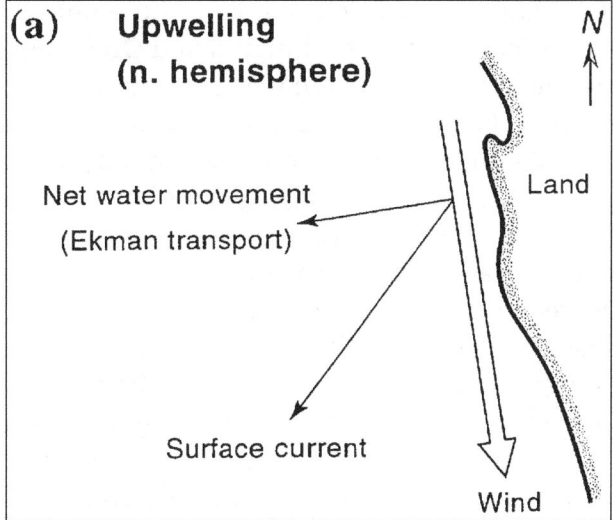

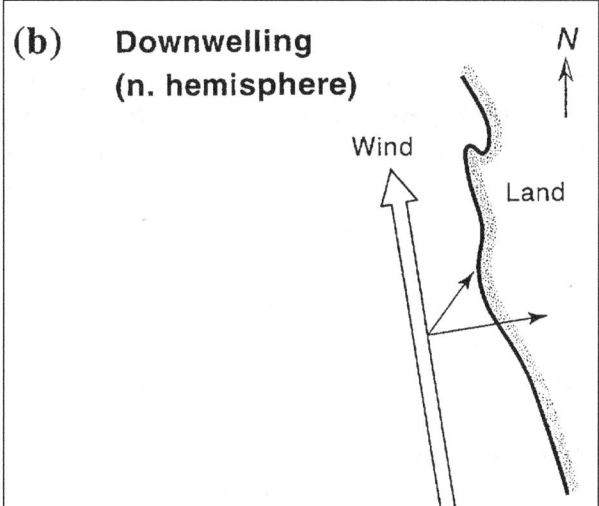

Figure: Schematic representation of coastal (a) upwelling and (b) downwelling in the northern hemisphere. The directions of the surface current and Ekman transport relative to wind direction would be reversed in the southern hemisphere, for both upwelling and downwelling events.

Coastal Upwelling and Downwelling

Surface winds which generally move surface currents 45° to the right (or left), and Ekman transport which averages 90° to the right (or left) of the prevailing wind direction, help establish significant circulation patterns when land masses "interfere with" or "disrupt" classical wind induced circulations unimpeded by land. A wind blowing from the north or northwest along a western coastline in the northern hemisphere, for example, will move surface water away from the coast at an approximate 45° angle, and will move water from the Ekman layer away from the coast at an angle of about 90°. Mass balance must take place, so water to replace that moved away from the coast must come from somewhere. It is brought up from a depth below the Ekman layer, in a process called coastal upwelling. Off the northwest coast of North America, as an example, the prevailing wind in spring and summer is from the NNW. Surface waters against the coast are thus moved offshore at this time and colder, nutrient-rich water from below upwells to replace that surface water moved offshore. In addition, the large Columbia River Estuary empties into this upwelling system between the states of Washington and Oregon in northwestern USA. Some years ago the Hanford Nuclear Site on the Columbia River discharged small amounts of several radionuclides into the river, and the river carried those nuclides into the coastal ocean. In spring and summer, the tongue of Columbia River water at sea could be identified on the surface by its signature of radioactivity. The tongue proceeded in a general south-southwesterly direction, about 45° to the right of the wind, as expected. Because these same NNW winds delivered non-radioactive, upwelled waters to the surface waters immediately adjacent to the coast, however, a fairly sharp gradient between the relatively warm, fresh, radioactive river water and the colder, saltier upwelled water was established; that is, the river water could not spread toward the coast at the surface, and the upwelled water could not spread seaward at the surface, creating a distinct frontal region. The effects given by the positions of land masses and by such things as river discharges into the oceans, can greatly contort general circulation patterns expected under a given wind regime. These contorting effects are particularly noticeable in the relatively shallow waters over continental shelves — and of course these are the regions having most direct impact on humans, and on which humans have the greatest impact.

During the autumn and winter, the prevailing winds off northwestern USA blow from the south. Coriolis Effect remains the same (to the right of the wind), so transport of surface and near surface waters is toward the coast (this is the time of year for beachcombing, as materials are brought onto the coast and often stranded there). At this time the Columbia River tongue, with its radioactive signature, is moved north and kept tightly against the coast of Washington state. Upwelling ceases because surface water is not moved away from the coast, but tends to pile up toward the coast, as mentioned previously. At this time there can be downwelling of water as some of that which tends to pile up at the coast must sink to maintain mass balance (some also moves laterally along the coast, intensifying a winter-time coastal current called the Davidson Current).

Convergence and Divergence Systems

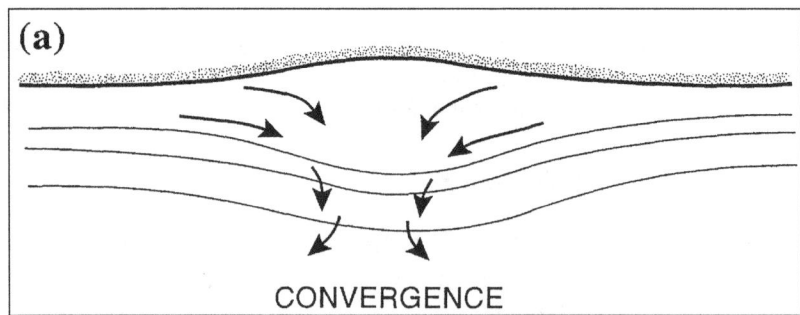

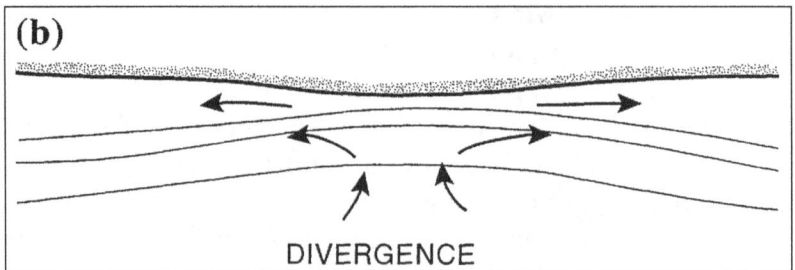

Figure: Schematic representation of oceanic (a) convergence and (b) divergence regions.

As water is transported to the right of the prevailing wind systems in the northern hemisphere, and to the left in the southern hemisphere on our rotating Earth, there is a tendency for surface water coming from opposing directions to pile up slightly in subtropical regions (particularly on the western sides of ocean basins in the sub-tropics), where winds are lighter than those in the polar regions. As the surface water tends to pile into a slight "hill", the pycnocline (a layer of strong density discontinuity below the surface) becomes depressed into a "valley". This creates a convergence. The most prominent convergences are the subtropical convergences between 20-40° N and S and the Antarctic convergence between 40-50°S. These mid-ocean convergences are very hard to detect, because the elevations of the water surface, and depressions of the pycnocline, are very small over large lateral distances.

Conversely, where water is moved away in opposite directions in a given region (by winds yielding Ekman transports in opposite directions), subsurface waters move to the surface to replace the surface waters that have been moved away, creating an oceanic divergence. It should be recognized that an oceanic divergence develops through a process similar to coastal upwelling, but with no coastal land barrier. Two prominent oceanic divergences are the Antarctic divergence (65-70°S) and the equatorial divergence created by the NE and SE Trade Winds imparting NW and SW Ekman transports to near-surface waters near the equator. Divergences are usually easier to spot than convergences because they often are rich with phytoplankton growth sustained by the nutrients being brought into surface waters from below. In this sense, open-ocean divergence zones again are similar to coastal upwelling regions.

Geostrophic Currents

So-called geostrophic currents can arise as water from the slightly elevated portions of the ocean (the "hills") tends to run down toward the more depressed regions (the "valleys") under the influence of gravity. However, the Coriolis Effect also is in play, so that the eventual current that ensues is a result of the balance between these two forces. The current does not run straight "downhill" into the "valley", as forced by gravity, nor does it run around the "hill" in one plane, under the sole influence of Coriolis: rather, it tends to describe a curving path down the "hill" toward the "valley". Satellite techniques are now used to accurately measure the sea-surface "hills" and "valleys", and thus more accurately map geostrophic currents in the world ocean.

Global Thermohaline Circulation and Water Mass Movement

A final type of global water movement, large-scale thermohaline circulation, needs to be addressed, because this, too, redistributes huge volumes of water and therefore redistributes all manner of dissolved and tiny paniculate matter (including radionuclides and pollutants). Global thermohaline circulation largely involves vertical water movements that initiate the horizontal movement of water from Polar Regions (particularly the North Atlantic and Antarctic Oceans). This circulation phenomenon controls temperature and salinity distribution in the deep ocean, and has a profound effect on distribution of other properties as well.

This massive vertical circulation is not caused by regional winds, but by density differences of different water masses. Density is controlled by temperature and salinity, with warm, fresh waters being less dense than cold, salty waters. When the density of water at the surface exceeds the density at depths below the surface (when surface water cools as a result of cooling air temperatures, for example), the water column becomes unstable and the more dense water sinks, displacing the less dense water immediately below it; hence a vertical circulation is set up. The water sinks until it reaches a depth of slightly more dense water, at which time the sinking mass moves laterally. On a global scale, particularly in the Atlantic Ocean near Greenland and in waters near Antarctica, large water masses of very low temperature and high salt content become very dense and sink quickly to depth. The Antarctic system provides a good example. Antarctic Bottom Water (AABW) is extremely cold (0-1° C) and salty enough so as not to freeze at those temperatures — it is the coldest water on Earth. It is slightly more dense than the cold North Atlantic Deep Water (NADW), so this latter water mass flows southward to replace, and override, the AABW. The AABW penetrates northward along the sea floor to about 45° S latitude before it loses its identity through wanning and mixing with other water masses.

Similarly, Antarctic Intermediate Water (AAIW) overrides the denser NADW, and the warmer surface sub-Antarctic water makes up the top layers. Enormous amounts of heat are transported into the system by the NADW, and lesser amounts are carried away from the Antarctic in AABW. Heat balance is generally maintained because huge quantities of heat are lost to the atmosphere at the Antarctic Convergence, with relatively small heat gains from solar radiation. The Antarctic system is thus an enormous heat exchanger for the global ocean, and the thermohaline circulation is also responsible for large movements of dissolved gases and other materials. Extensive mixing with waters in the West Wind Drift around Antarctica (the Antarctic Circumpolar Current) allows large volumes of Antarctic water to be carried into the Pacific Ocean. The Pacific itself is not so intensely charged with cold, dense water as the Atlantic; as North Pacific waters are less salty than North Atlantic Waters and so do not sink as deeply as those in the North Atlantic. Nor is the southerly excursion of the densest North Pacific water as prominent as that of the North Atlantic. North Atlantic Deep Water reaches the Antarctic region, whereas sinking North Pacific water does not reach the equator. Knowledge of these vast, deep circulation patterns is crucial to understanding potential effects and hazards of deep-sea waste disposal.

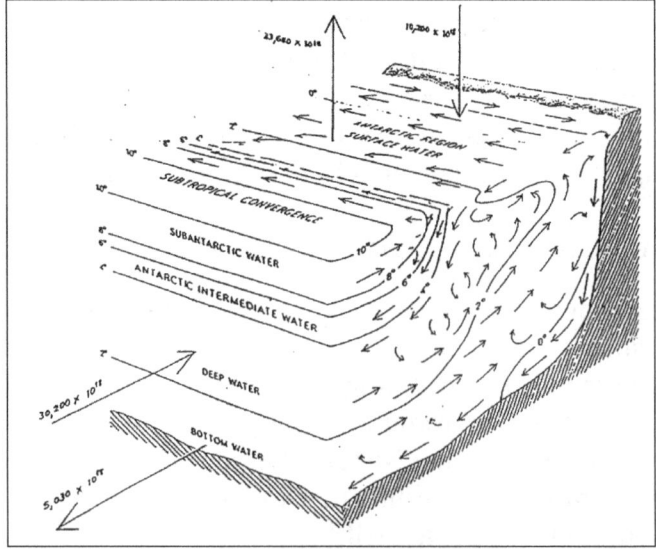

Figure shows schematic representation of the complex current flow in the Southern Ocean. Figures at the left are estimates of the amount of heat, in calories, transported annually by the deep and bottom waters. Figures at the top show that the annual heat given off to the atmosphere by the Southern Ocean greatly exceeds the heat received from the sun.

Coastal and Near-surface Processes

Significance of Coastal Processes

Coastal processes often become relatively more significant than open-ocean processes because mankind is most closely associated with coastal environments, through fishing, recreational pursuits, transportation routing, mineral extraction (including oil and gas), and dumping of wastes.

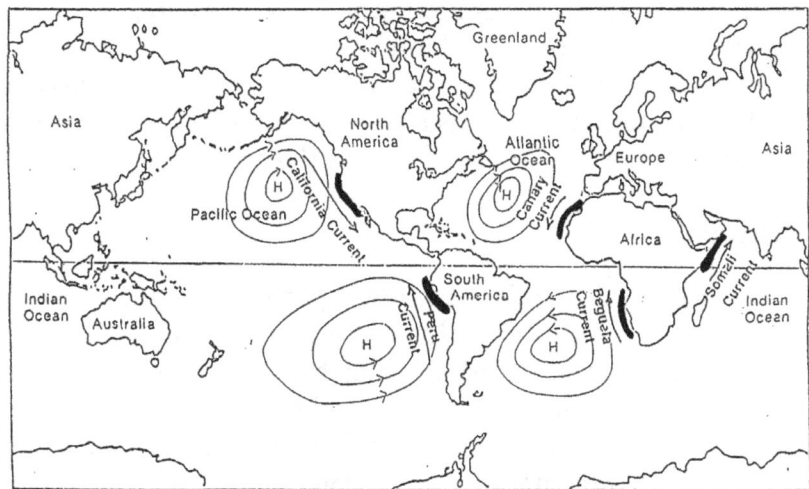

Figure: Major coastal upwelling areas of the world, and the atmospheric pressure systems and prevailing oceanic currents associated with them.

Significance of Coastal Upwelling Systems

Coastal upwelling has already been introduced, but it clearly is one of the most significant coastal processes and deserves further treatment. Major upwelling areas of the world ocean represent only about 0.1% of the ocean's surface area, but perhaps 50% of its fish production. Coastal upwelling areas are characteristically associated with eastern boundary currents (western sides of continents in the Atlantic and Pacific Oceans), although the upwelling zone off the Somali coast is an exception, a product of the SW (summer) monsoon system. More minor or more ephemeral upwelling systems can be found anywhere that winds, in concert with the Coriolis effect, move surface waters away from a coastline (with subsequent replacement of those waters with waters upwelled from greater depths). The most persistent systems are those in which the prevailing winds are steady enough over long enough time periods to keep the upwelling process active for long periods. Long-term upwelling continues to resupply surface waters with dissolved nutrients from below the Ekman layer, thus providing the fertilizer necessary to sustain luxurious phytoplankton growth, which in turn supports food webs stable enough to allow large production of fish biomass. Upwelling areas in the westerly wind belts (western North America, northwest and southwest Africa) are seasonal, as the atmospheric high pressure systems generating the upwelling-favorable winds themselves are seasonal in position, and the Somali upwelling system is seasonal because of the seasonally of the monsoon winds, as noted earlier. Within any given season upwelling-favorable winds may disappear for short periods, thus turning off the upwelling circulation with its supply of nutrients to the euphotic zone; however, upwelling-favorable winds are the predominant winds during the summer seasons in the westerly wind belts, so that upwelling circulation is persistent enough to maintain a productive food web. The most productive upwelling system is that off northwestern

South America, principally Peru and Ecuador. This system is driven by the Trade Winds, which, although relatively light, are nevertheless persistent throughout the year. This system boasts some of the highest fish production in the world, although overfishing and El Nino events have reduced stocks to very low levels in recent years.

The major upwelling areas have somewhat different patterns of distribution of nutrients, and therefore different patterns of phytoplankton growth and food web structure. These different patterns are brought about by the different configurations of bathymetry and coastal topography in the different areas, as well as by different wind intensities and persistence. It is clear that disposal of radioactive or other wastes into any upwelling system, or locating processing plants in an upwelling region, would be harmful. The systems are relatively shallow, very productive, and would tend to redistribute near-bottom materials back toward the coast.

Tides and Tidal Currents

Coastal areas also experience tides and tidal currents, capable of locally redistributing materials injected into estuarine or near shore waters. Tide-generating forces include gravitational attraction between the Earth and moon, centrifugal forces created by the rotation of the Earth and moon, and gravitational forces between the Earth/moon system and the sun. A parcel of water on the Earth's surface nearest the moon is about 3% closer to the moon than a parcel on the Earth's surface farthest from the moon; hence, the moon's gravitational attraction is greatest on the water nearest the moon, and least on the water farthest from the moon. Centrifugal forces are equal over the Earth's surface, however. Thus, on the side nearest the moon, the gravitational attraction of the moon exceeds the centrifugal force, so that the water is pulled out toward the moon. On the side furthest from the moon, the centrifugal force exceeds the gravitational pull, so that there, too, the water is pulled away from the earth. A slightly oval water envelope is thus created around the sphere of the Earth in this idealized scenario with the bulge of the oval on both sides of the Earth being the high tides, and the slightly flattened sides of the oval being the low tides. Remember that the earth revolves under its water covering, so that the tidal bulge (high tide), to someone standing at a fixed point on the revolving Earth, seems to appear (increase) and disappear (decrease) over a fixed period of time. If, for example, the moon is in the plane of the Earth's equator, the two tidal bulges on opposite sides of the Earth will be centered at the equator. After the Earth rotates 90°, the same point on the equator would register a low tide. Continual 90° rotations yield successive; equal-sized high and low tides from the fixed point - two high tides and two low tides over a period slightly greater than one full day.

The moon does not maintain a fixed position relative to the Earth, however; i.e., the moon is not always in the plane of the equator, creating tides of equal magnitude each tidal day. Therefore, a person at a fixed position on Earth would see two high tides and two low tides during each tidal day, as before, but over a period of successive days the height of the high tides and depth of the low tides would be different as the position of the moon changed relative to a fixed point on the Earth; thus, higher high tides and lower low tides are registered relative to other times.

The sun also exerts a gravitational effect on the Earth's water envelope, but it is more than 50% weaker than that of the moon because of the enormous distance between the Earth and the sun. Also, solar tides have a 12 hr. period (the time between two high tides, or between two low tides), while the lunar period is about 12.5 hours. Finally, the Earth takes a full year to cycle the sun, whereas the moon cycles the Earth once per month.

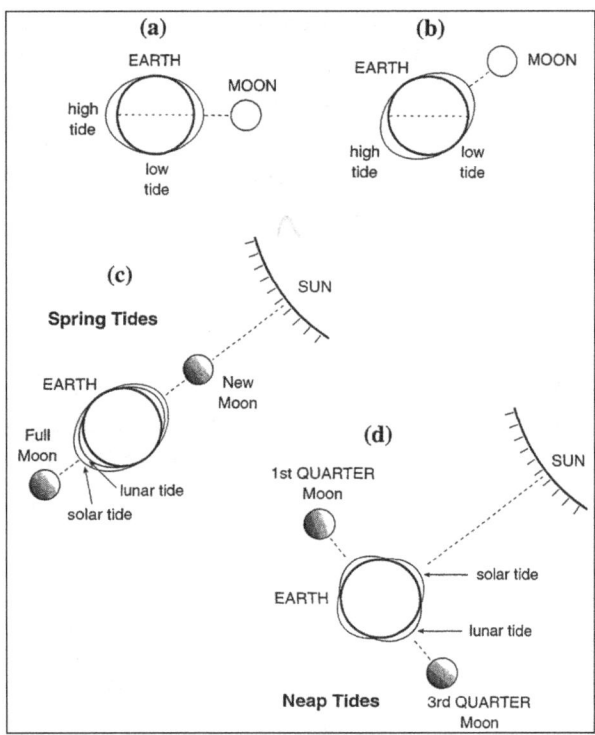

Figure: Schematic representation showing how tides are created.

Solar and lunar tides acting together or in opposition create spring tides or neap tides, respectively. Spring tides, which occur approximately every two weeks around the times of the new moon and full moon, occur when the sun and moon line up with the Earth, so that their tide-generating forces act together to create extra-large bulges in the tidal envelope (i.e., extra high tides). Conversely, during the first and third quarters of the moon, the solar and lunar tide-generating forces partially counteract each other by acting in different directions, resulting in the lowest tidal range - the neap tides. There are further complexities to tides than the above so-called "equilibrium theory".

The tide in every marginal sea, bay, harbor, or estuary is greatly influenced not only by the magnitude of the ocean tide at the mouth, but also by the natural period of the basin and the cross-section of the mouth. The small tidal range in the Mediterranean Sea (about 0.5 meter), for example, can be explained by the small opening through the Strait of Gibraltar into the Atlantic, relative to the size and depth of the Mediterranean basin. The effect of tides on distributing radionuclides in the Mediterranean is vanishingly small. On the other hand, in the Bay of Fundy (New Brunswick, Canada), where the natural period of the basin is near the tidal period (12 hours), large standing waves are set up, which course back and forth through the basin giving an extreme tidal range (ranges on the spring tide are about 15 m in the Bay of Fundy). Movement of materials by these tidal bores is extremely great. Thus, tidal effects in coastal areas can be either great or small, alternately exposing and covering great areas of beach or shoreline, or not making much difference at all. Knowledge of the tidal features in a given area of coast or estuary is exceedingly important when considering radionuclide waste discharge or other discharges.

Nearshore Surface Waves, Internal Waves and Surf

Surface waves are not particularly significant in distribution of materials until they crest over to form surf adjacent to beaches. Surf is lateral movement of surface waters, and thus can distribute materials onto beaches. Internal waves, however, can redistribute much heat, dissolved gases, and other materials in the ocean's interior. In principle they are like surface waves, but they are usually found at the interface of water layers of different density. Internal waves can be much higher, longer, and move much more slowly (with longer period) than surface waves, because the density differences between two water layers are much less than between water and air.

Standing Waves and Tsunamis

At times the whole water mass in an enclosed basin (e.g., the Red Sea or the Bay of Fundy) can slosh back and forth, much as water in a bathtub can be made to slosh back and forth by applying and releasing some pressure on the water at one end of the tub. An earthquake or some other major phenomenon can often set off these standing waves in a natural basin. The edges of a standing wave at the shore alternately expose and inundate large areas of shoreline, thus potentially redistributing near-shore materials. The ultimate standing waves are seismic sea waves, or tsunamis, which can be devastating when they occur; however, they are rare. Tsunamis are usually caused by large earthquakes beneath the sea floor which act to accelerate a wall of water in some direction. The tsunami is similar to a shallow-water standing wave, but originates in deep water and thus can transport an enormous volume of water very rapidly (up to 800 km/hr). Tsunamis are often undetectable by ships at sea, and only wreak havoc when they encounter a land mass.

Small-scale Thermohaline Circulation

Thermohaline circulation, on a smaller vertical scale than that addressed earlier, is responsible for much water movement and distribution of properties in the coastal ocean and in upper layers of the open temperate ocean.

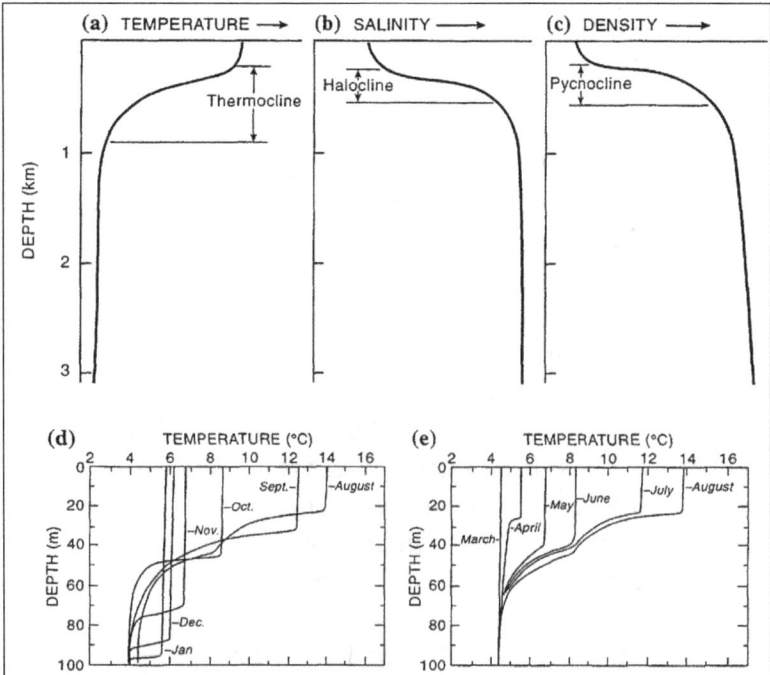

Figure: Schematic representation of a thermocline (a), halocline (b), pycnocline (c), fall-winter decay of a thermocline (d), and spring-summer establishment of a thermocline (e).

Because radionuclides and most other materials enter the ocean through the surface layers, these upper-water circulations are very significant. The circulations are generated when stable, near-surface water columns are disrupted (usually seasonally due to seasonal changes in temperature and salinity of those waters). Relatively stable upper-water columns occur when surface waters, warmed by the sun, form a relatively thin mixed layer over a region of fairly rapid temperature change with depth. This layer of temperature discontinuity is the thermocline. Below the thermocline temperatures remain cold, and change very little all the way to the sea floor. Sometimes salt is the principal agent of stability of near-surface waters in the ocean, so that a shallow region of low-salinity water overrides a zone of maximum salinity change with depth (the halocline), which in turn overrides deep waters with high salinity that changes very little with depth. The combined effects of temperature and salinity are responsible for the density of the water, so that a stable water column features a shallow surface layer of low-density water (relatively warm and fresh) which overrides a zone of maximum density change with depth (the pycnocline), which in turn overrides the densest water (relatively cold and salty). Waters of different densities resist mixing, so that any radionuclides put into surface waters overriding a well-established pycnocline would remain in those surface waters for a long time, if physical processes were the only processes involved in radionuclide distribution.

In tropical waters, upper water columns can remain relatively stable for long periods of time, the stability mainly being upset by episodic events such as storms at sea, or by internal waves along interfaces between the pycnocline and waters above or below it. In temperate seas, winter lowering of atmospheric temperatures, and/or rainfall and river runoff in the coastal ocean, causes the water-column stability of summer to break down. As autumn air begins to cool the immediate water surface layer, that water sinks until it reaches water of equal density. Further cooling of the air acts to cool more surface water, which also sinks. This process continues, setting up a vertical convection of cooler water sinking through the slightly warmer water until the upper layers become equally dense and therefore capable of mixing easily. As spring approaches, the progression reverses; i.e., the winter water column begins to warm at the surface. This warmer surface layer resists mixing with the slightly colder water just below it, initiating the formation of a thermocline. As the surface waters continue to warm, the thermocline deepens until the incoming summer radiation attains its maximum possible effect. The progression then begins to reverse again. It should be noted that these thermohaline circulations are identical to global thermohaline circulations involving mass transports of deep water, except the former are near-surface phenomena usually occurring on a seasonal basis in temperate and subpolar seas. Seasonal thermohaline circulations rarely involve depths beyond 100 m or so, but note that this depth zone is the approximate productive zone of the sea and the one contiguous with the atmosphere and with river and coastal discharges.

Small-scale Processes

Scaling for Sampling Protocols

When studying oceanic water movements relative to distributions of radionuclides, the tendency is always to present the large-scale processes and ignore the small-scale processes. However, from the standpoint of knowing how to sample, when to sample, and how often to sample water for valid distributional analysis of radioactivity at sea, it is imperative to have some understanding of time and space scales of physical processes (such as mixing and diffusion) relative to biological and radiochemical processes of interest.

Simple formulae can often be used to determine characteristic scales; for example,

$$T = \frac{(q)}{\left(\dfrac{dq}{dt}\right)} = \text{characteristic time (T) for sampling quantity (q)},$$

Where, q = concentration of a long-lived radionuclide in the water, for example, and dq/dt = the time rate of change of quantity q in that water (e.g., $\mu g/m3 \div \mu g/m3/hr$ = hrs). T is the number of hours one has to sample q before q is dissipated to generally unmeasurable levels by dispersion, sinking, biological uptake, or other mechanisms (presuming q is a one-time input to the sampling region). If q is a radionuclide with short half-life, then radioactive decay must also be accounted for. Knowledge of space scales required for sampling are also important, and this involves selection of the numbers and spacing of sampling sites to adequately represent the spatial domain of interest.

Reynolds Number

One dimensionless number that has proved useful in delineating time and space scales significant for a particular sampling scheme is the Reynolds number. This number gives the relationship between turbulent processes and molecular processes in a fluid, and is defined as:

$$Re = \frac{ul}{V}$$

Where, ul=a characteristic length scale (l) times a characteristic velocity (u), of water, a particle, or an organism; and V= kinematic viscosity of the fluid (for water, kinematic viscosity = viscosity/density $\approx 10^{-2}\,cm^2\,sec^{-1}$, a measure of molecular motions). A low Re represents more viscous, laminar flow, while a high Re represents turbulent flow. A one-micron-sized particle of fallout radioactivity, for example, operates in a very low-Re environment (very tiny "length" scale and very weak sinking velocity relative to the kinematic viscosity of the water in which it was deposited). A zooplankton organism 1mm in length also operates in a relatively low-Re environment (the short length and slow swimming speed of the zooplankter relative to the kinematic viscosity of the water in effect gives the same mobility problem to the zooplankter as a human being would have trying to swim through thick molasses). On the other hand, a full-grown tuna lives in a high-Re environment (long length and rapid swimming ability of the fish relative to the kinematic viscosity of the water). Clearly one would not design a similar sampling program for fallout particles, zooplankton and tuna; however, would one design the same program for 1μm and 100μm particles (e.g., fallout particles and diatoms), for 10mm zooplankton and 10mm zooplankton. Small-scale processes are important, and correct sampling scales are required to adequately assess them.

Richardson Number

The Richardson number (R_i) is another dimensionless number, relating a stabilizing buoyancy to a destabilizing shear. This measure is important in determining when a stable upper water column turns into a mixed water column, and to what extent:

$$R_i = \left[\frac{\left(-\dfrac{g}{e} \right) \left(\dfrac{de}{dz} \right)}{\left(\dfrac{du}{dz} \right)^2} \right]$$

Where, the numerator is a measure of buoyancy (a sink for turbulent energy, the so-called Brunt-Vaisala number), and the denominator is a measure of shear (a source of turbulent energy). In the formula, g = acceleration due to gravity, e = density of the water, u = velocity, and z = depth. Richardson numbers between zero and one denote the region where shear overcomes stability, leading to a turbulent condition (0.25 is approximately the balance point where neither shear nor buoyancy controls). An R_i greater than one is a stable water column (a greater sink than source for turbulent energy).

Branches of Oceanography

Physical Oceanography

Physical oceanography is a branch of oceanography that deals with the physical and chemical properties of ocean water and the topography and composition of the ocean bottom. The physical properties of seawater include both 'thermodynamic properties' like density and freezing point, as well as 'transport properties' like the electrical conductivity and viscosity. Density in particular is an important property in ocean science because small spatial changes in density result in spatial variations in pressure at a given depth, which in turn drive the ocean circulation.

Physical properties can be measured directly. However, direct measurements can be complicated to carry out, especially in the field, and in many cases it is more convenient to measure a few important 'state variables' on which the properties depend.

Physical properties vary with the amount of heat and the amount of dissolved matter contained in the water, as well as the ambient pressure. Important state variables measured for parcels of water in the ocean are therefore temperature, which is related to the heat content, salinity, which is related to the amount of dissolved matter, and the pressure. In addition to controlling physical properties, the variation in space and time of temperature and salinity are also important water mass tracers that can be used to map the ocean circulation. Density is usually calculated using a mathematical function of temperature, salinity, and pressure, sometimes called an equation of state. For many years the internationally accepted standard for seawater densities has been the 1980 International Equation of State, known by the acronym EOS-80. However, a new international standard for seawater density, as well as all other thermodynamic properties, has recently been developed. This new standard is called the Thermodynamic Equation of Seawater (2010), or TEOS-10. In turn, standards like EOS-80 or TEOS-10 rely on other international standards that precisely define the state variables of temperature and salinity. Standards thus play an important role in ocean science.

Table: Ocean state variables, their typical ranges and mean values in the ocean, and the accuracy to which they are measured (or estimated) in the deep ocean.

Variable	Ocean Range	Ocean Mean	Required Accuracy
Temperature	-2°C to 40°C	3.5°C	±0.002°C
Absolute Salinity	0 g/kg to 42 g/kg	34.9 g/kg	±0.002 g/kg
Pressure	0 dbar to 11000 dbar	1850 dbar	< ±3 dbar
In-situ Density	1000 to 1060 kg/m³	1036 kg/m³	±0.004 kg/m³
Pressures in the ocean are usually measured in dbar (decibars), with values offset to read zero at the surface. Pressure values expressed in this way are numerically similar (within 2%) to depth in meters below the surface. That is, the pressure at a depth of 1000 meters is about 1000 dbars.			

Temperature

Objects in contact with one another will tend toward thermal equilibrium by an exchange of heat between them. Thus, if we know the temperature of one object (call it a thermometer), and it is in thermal equilibrium with water around it, which will occur after enough time has passed, we also know the temperature of the water. Conversely, if water and a thermometer within it are at different temperatures, then there must be a flow of energy (heat) between the water and the thermometer. This gives us both a way of talking about energy and a way of measuring temperature using a known reference.

In turn, the temperature of the thermometer can be related to its physical properties. These physical properties include the volume-to-mass ratio of a solid, liquid, or gas, or the electrical resistance of a metal or a semiconductor. Temperature can therefore be determined through direct measurements of these properties.

The simplest way is to define two reference points, plus a method that can be used to interpolate between them. In 1742, Anders Celsius defined a temperature scale in which the freezing point of water (at sea-level pressure) was taken as a lower reference point with a value of 0, with the difference between the freezing point and the boiling point (also at sea level pressure) taken as 100 units, measured in terms of the change in volume of a fluid.

As time went on, more and more precisely specified definitions (or standards) for temperature became necessary as scientific questions required inter-comparing more and more accurate observations of temperature. However, in order to preserve a historical continuity, these redefinitions were usually formulated in such a way as to allow easy comparison with older observations. We therefore still use the °C as a unit of temperature, and water still freezes at about 0°C and boils at about 100°C, although temperature scales are no longer defined in terms of the freezing and boiling points of water.

An important step in the history of temperature occurred in the 1800s, when the concept of temperature gained a theoretical backing with the development of the science of thermodynamics. William Thomson (who later became Lord Kelvin) suggested a thermodynamic temperature scale in which the ratio of temperatures above an 'absolute zero' would be in proportion to the heat absorbed and rejected by a theoretical construct called a Carnot cycle engine, operating between these two temperatures.

A thermodynamic scale is thus implicitly referenced to a lower limit of absolute zero, which is defined to be 0K (kelvin). The freezing point of water was chosen to be a second reference. By numerically defining the temperature of the freezing point of water to be 273.15K, the thermodynamic temperature of the boiling point is found to be 373.15K, 100 units higher, so that an interval of 1K and 1°C are identical.

However, determining thermodynamic temperature is very difficult, and for most applications more practical methods are needed. These practical definitions of temperature generally rely on fixing the values of certain reference points based on careful measurements of their thermodynamic temperatures, and then specifying a way of interpolating between them. Points at which different phases of particular materials co-exist are useful reference points. However, rather than boiling and freezing points (which depend on the ambient pressure), better fixed points occur at the triple points of different substances. A triple point is the single unique combination of temperature and pressure at which solid, liquid, and gas phases of a particular substance can all coexist.

The current standard for temperature is the International Temperature Scale, or ITS-90. In ITS-90, temperatures in the range of oceanographic interest are set by:

- The triple point of mercury, defined to be $\equiv$ -38.8344°C exactly.

- The triple point of pure water with a specified isotopic composition, defined to be $\equiv$ 0.01°C (or 273.16K) exactly.

- The melting point of gallium, defined to be $\equiv$ 29.7646°C.

In addition, a method of calculating temperatures between these reference points is also described by ITS-90. Such temperatures are interpolated using a specified polynomial function of the measured electrical resistance of a platinum wire. Temperatures outside the oceanographic range are defined using other fixed points and methods of interpolation.

Temperatures specified using the previous international standard temperature scale (the International Practical Temperature Scale 1968 or IPTS-68) differ from ITS-90 by up to 0.01°C over the range of oceanographic interest. Although our normal day-to-day activities usually don't require accuracies of more than about 0.1°C at best, a difference of 0.01°C is somewhat larger than the accuracy to which deep ocean measurements are now made. Careful adherence to standards is therefore necessary to allow us compare measurements made by different people around the world, and to compare contemporary measurements to measurements made in the past and in the future.

Although in-situ temperature t defined by ITS-90 is generally the quantity we measure, it is not the most useful variable for describing heat content itself. Two effects can cause problems. First, the energy required to change the temperature of seawater by a fixed amount (say, 1°C), called the heat capacity, is itself a function of temperature and salinity. It takes about 5% less energy to heat average seawater by 1° C than it does to heat the same mass of freshwater by 1°C. Second, the effects of pressure can act to change the in-situ temperature of water without changing the heat content.

Squeezing typical seawater (or air) causes the temperature to rise. A pressure of 100 atmospheres (or about 1000 dbar) is enough to increase measured seawater temperatures by about 0.1°C. However, the temperature of near-freezing fresh water actually falls as pressure increases.

To account for the pressure effects, a variable called potential temperature, denoted θ, is traditionally used in oceanography. The potential temperature of a water parcel is the temperature that would be measured if the water parcel were enclosed in a bag (to prevent the loss or gain of any salt) and brought to the ocean surface adiabatically (i.e., without exchanging any heat with its surroundings). The potential temperature is therefore insensitive to pressure by definition, but is lower than the in-situ temperature by about 0.1°C for every 1000 m of depth increase.

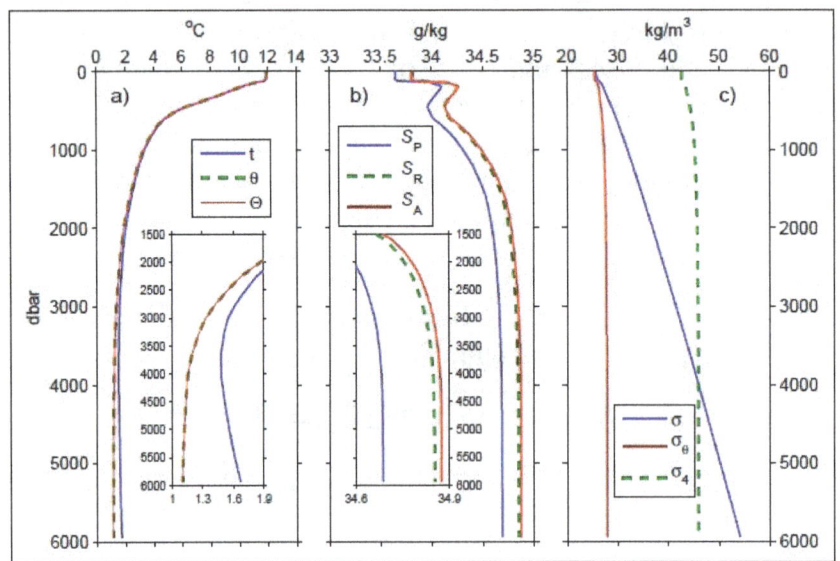

Figure: A vertical profile of temperature and salinity at 39°N, 152°W in the North Pacific.

Figure shows (a) Vertical profiles of in-situ temperature t, potential temperature θ, and Conservative Temperature Θ. Inset shows an expanded view of deep ocean values. Potential and Conservative temperature are within 0.0003°C in the deep ocean. Note that heat content (proportional to Θ) actually decreases with depth, but pressure effects cause a rise in the in-situ temperature t. (b) Vertical profiles of Practical Salinity S_P (with no units), Reference Salinity S_R, and Absolute Salinity S_A. (c) Vertical profiles of in-situ density (σ), potential density referenced to the sea surface (σ_θ), and potential density referenced to 4000 dbar (σ_4). Most of the in-situ density increase with depth is due to pressure effects, which are removed in the calculation of potential densities, showing how weakly stratified the ocean actually is.

However, potential temperature does not account for the varying heat capacity of seawater, and is therefore not a conservative measure of heat content. A variable is conservative if its value in a mixture of two parcels of seawater is the average of its values in the two initial parcels. Conservative variables are very useful in making budgets. Although the heat content of a mixture of two seawaters is the average heat content of the two, both the in-situ and potential temperature of a mixture can differ from the average temperature of the two initial parcels.

A different type of temperature, defined in TEOS-10, more precisely scales with heat content and is for all practical purposes conservative, as well as insensitive to pressure. It is called Conservative

Temperature, and denoted Θ. Conservative Temperature differs from potential temperature by as much as 1°C for warm fresh waters, but is usually well within ±0.05°C for most ocean waters.

Although the differences between these different types of temperature are small, they can be significantly larger than the precision to which temperature measurements of the ocean are routinely reported. It is therefore important to make clear which temperature is under consideration in any discussion.

Salinity

Salinity is a measure of the 'saltiness' of seawater, or more precisely the amount of dissolved matter within seawater. Operationally, dissolved matter is that which remains after passing the seawater through a very fine filter to remove particulate matter. Historically, a glass fiber filter with a nominal pore size of 0.45 μm was used. More recently, filters with 0.2 μm pores have become standard, since filters with this pore size will catch the smallest bacteria.

However, the history of the salinity concept and its various definitions (which have changed over time) is a long and complex story, dating back to the late 19th century. The story is complex for two reasons. First, any useful definition of salinity contains approximations of some kind. These approximations are necessary because the dissolved matter in seawater is a complicated mixture of virtually every known element and it is impossible to measure the complete composition of every water sample. Second, subtle technical details of these approximations, which have undergone changes as more has been learned about seawater, are very important in practice. These details are important because the required measurement accuracy for salinity, necessary to understand the ocean general circulation, is extremely high (about ±0.006%), so that even small changes in numerical values can have significant implications if incorrectly interpreted.

Most useful definitions of salinity are rooted in the well-known fact that the relative ratios of most of the important constituents of seawater are approximately constant in the ocean (the Principle of Constant Proportions). Therefore, practical but approximate measures of the total dissolved content can be found by scaling measurements of a single property.

Originally the property most conveniently measured was the chlorinity, or halide ion (mainly Cl⁻ and Br⁻) concentration. Chlorinity was measured using a straightforward chemical titration, and then converted to a measure of salinity using a simple linear function. Such salinities can often be identified by an attached unit of ppt, or the symbol ‰.

However, almost all modern estimates of salinity rely on measurements of the electrical conductivity (or, at high precision, on measurements of the ratio of the conductivity of a sample of seawater to the conductivity of a special reference material called IAPSO Standard Seawater). Since the electrical conductivity of seawater is also highly temperature-dependent, and mildly pressure-dependent, temperature and pressure must also be measured in this approach. The conversion from measured temperature, pressure, and conductivity to salinity is complex and nonlinear. Since the early 1980s, oceanographers have used a calculated value formally called the Practical Salinity (denoted S_p) as a proxy for true salinity. Practical Salinity is defined as a function of temperature, pressure, and conductivity by another standard, the Practical Salinity Scale 1978 (or PSS-78). When oceanographers use the word salinity they often mean Practical Salinity, although it is better to use the full name to prevent ambiguity.

It is important to emphasize that Practical Salinities do not have units. This fact, confusing to non-specialists, is related to technical issues that prevented an absolute definition when PSS-78 was constructed. Sometimes this lack of units is awkwardly handled by appending the acronym PSU (Practical Salinity Units) to the numerical value, although doing so is formally incorrect and strongly discouraged. Practical Salinities are numerically smaller by about 0.5% than the mass fraction of dissolved matter when this mass fraction is expressed as grams of solute per kilogram of seawater. Practical Salinities were, however, defined to be reasonably comparable with numerical values of chlorinity-based salinities, to maintain a historical continuity.

The special reference material used to calibrate salinity instruments, IAPSO Standard Seawater, is manufactured by a single company (Ocean Scientific International Ltd., UK) and is created using seawater obtained from a particular region of the North Atlantic. Although the use of Standard Seawater to determine Practical Salinity has been routine for many years, the dependence of Practical Salinity measurements on a physical artifact known to degrade with age leads to a number of technical problems, especially in terms of the long-term stability and inter-comparability of high-precision ocean measurements.

The new seawater standard TEOS-10 defines a better measure of the salinity, called Absolute Salinity (denoted S_A). This new definition incorporates several features designed to address the technical difficulties and provides the best available estimate of the mass fraction of dissolved matter. It is usually associated with an attached unit of g/kg.

First, the definition of salinity is no longer based on properties of IAPSO Standard Seawater. Instead, the best estimates of the concentrations of the important inorganic components of Standard Seawater are used in TEOS-10 to exactly define artificial seawater with Reference Composition. For practical and historical reasons, the definition of Reference Composition ignores dissolved organic matter, as well as most gases, although it otherwise includes the most important constituents in real low-nutrient seawater.

Table: Reference Composition of seawater with $S_P \equiv 35.000$ and $S_R \equiv 35.16504$ g/kg.

Reference Composition	mmol/kg	mg/kg
Na^+	468.9675	10781.45
Mg^{2+}	52.8170	1283.72
Ca^{2+}	10.2820	412.08
K^+	10.2077	399.10
Sr^{2+}	0.0907	7.94
Cl^-	545.8695	19352.71
SO_4^{2-}	28.2353	2712.35
Br^-	0.8421	67.29
F^-	0.0683	1.30
HCO_3^-	1.7178	104.81
CO_3^{2-}	0.2389	14.34
$B(OH)_3$	0.3143	19.43
$B(OH)_4^-$	0.1008	7.94

CO_2	0.0097	0.43
OH^-	0.0080	0.14
Observed Variations seen in real seawater		
O_2	0 - 0.3	0 - 10
N_2	0.4	14
$Si(OH)_4$	0 - 0.17	0 - 16
NO_3^-	0 - 0.04	0 - 2
PO_4^-	0 - 0.003	0 - 0.2
ΔCa^+	0 - 0.1	0 - 4
ΔHCO_3^-	0 - 0.3	0 - 20
Dissolved Organic Matter (DOM)	–	0 - 2

Concentrations in seawater of higher or lower salinities can be found approximately by scaling all values up or down by the same factor. Units of concentration are per kilogram of seawater. Real seawater contains additional constituents which are not included in the Reference Composition but whose concentrations (and their variation) may be larger than 1 mg/kg. Concentrations of these constituents do not increase or decrease with salinity but are largely controlled by biogeochemical processes.

Next, a numerical Reference Salinity (denoted S_R) is defined, representing the mass fraction of solute in this Reference Composition seawater. Reference Salinity has units of grams of solute per kilogram of seawater, and is numerically determined by multiplying the concentrations of the different components of the Reference Composition by their atomic weights, and then summing. Salinities defined in this way are said to lie on the Reference Composition Salinity Scale. Note that uncertainty in the atomic weights themselves contributes an uncertainty of about 1 mg/kg to this definition.

Standard Seawater is now treated as a physical artifact that approximates Reference Composition Seawater. A particular sample of Standard Seawater is then assigned a Reference Salinity on the Reference Composition Salinity Scale. This Reference Salinity is numerically different from the Practical Salinity of the sample, but it can be obtained from the conductivity-based Practical Salinity using a simple scaling. However, Reference Salinity can also be estimated using other approaches (e.g., by direct measurements of density and inversion of the TEOS-10 equation of state).

Although the definition of the Reference Composition provides a standard for the definition of the salinity of Standard Seawater, when considering real seawaters an additional problem arises. This is because the relative chemical composition of seawater is in fact slightly different in different geographic locations. The most important variations that occur in the real ocean arise from changes in the carbon system, and in the concentrations of Calcium (Ca^{2+}) and the macronutrients nitrate (NO_3^-) and silicic acid ($Si(OH)_4$). These constituents are affected by biogeochemical processes in the ocean. They are removed by the formation of biological material, and returned by its dissolution.

When using PSS-78 these changes in relative composition are ignored. However, this means that waters of the same Practical Salinity, from different parts of the ocean, may contain different mass fractions of dissolved matter. The difference can be as large as 0.025 g/kg in the open ocean. In coastal waters, where the presence of river salts is an additional factor, the difference can be as large as 0.1 g/kg. Differences of this size are more than an order of magnitude larger than the precision to which salinity is reported.

Under TEOS-10 these changes in the relative composition are explicitly accounted for in the definition of Absolute Salinity. The TEOS-10 Absolute Salinity can be determined by first measuring the electrical conductivity, temperature, and pressure of a water parcel, as before. Then Reference Salinity is calculated as if the water had Reference Composition. Finally, a small correction factor is added to account for the compositional variations. This correction, also known as the salinity anomaly, is denoted ΔS_A. It is roughly correlated with the concentration of macronutrients in seawater, and is largest in the deep North Pacific, where these concentrations are greatest.

Density

The most important thermodynamic property of seawater for studies of oceanic circulation is its density (denoted ρ). Typical densities span a narrow range. It is therefore conventional in discussions of the ocean to use a derived variable σ for density where,

$$\sigma = \rho / \ (\text{kg/m}^3) - 1000$$

So that the leading '10', virtually always present in numerical values of density, is dropped. For example, a density of 1027.534 kg/m³ would usually be written as a σ-value of 27.534 kg/m³.

Density depends on heat content and salinity. Since seawater is not perfectly incompressible, it also varies slightly with pressure. The variation of σ with Conservative Temperature and Absolute Salinity can be illustrated using a T-S diagram, or more properly a Θ-S_A diagram. A water parcel with a particular Θ and S_A is plotted as a point in such a diagram. By calculating the density at all possible points (keeping pressure fixed at a specified value), and contouring the calculated values, isopycnal lines can be drawn on the diagram joining together different combinations of temperature and salinity that result in the same density.

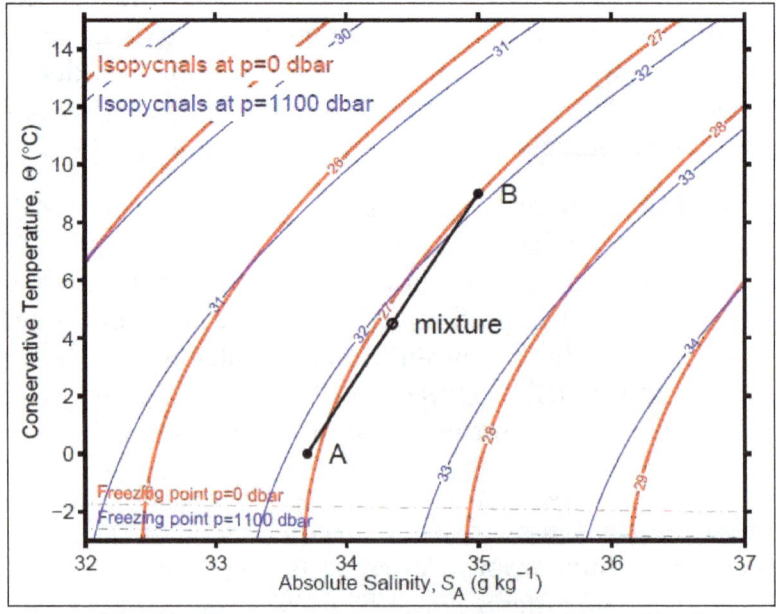

Figure: This Θ -S_A diagram can be used to illustrate relationships between temperature, salinity, and density for different water masses.

Lines of constant σ are contoured at the surface (red curves for p = 0 dbar), and at about 1100 m depth (blue curves, p = 1100 dbar). Near the surface water masses A and B, with different temperatures and Absolute Salinities, both have σ ≈ 27 kg/m3, with the density of water A slightly less than the density of water B. However, at 1100 dbar where σ ≈ 32 kg/m3 water mass A is denser than water mass B.

As one would expect, more saline waters at a particular temperature are more dense. Increasing the Absolute Salinity by about 1 g/kg increases density by about 0.8 kg/m³. However, isopycnal lines joining water types of different temperature and salinity but the same density are not straight. Instead, they are curved, with a right-facing concavity. The curvature of the isopycnals shows that the effects of temperature on density are very much greater in warm waters (i.e., near the surface in the tropics) than they are in cold waters found at depth and in the Polar Regions.

However, there is always a small temperature effect, and so the isopycnal curves never become vertical for seawaters with salinities typical of the open ocean. Although maximum densities occur at temperatures of around 4°C for fresh waters, for Absolute Salinities greater than 23.8 g/kg, seawaters at the freezing point are most dense. Freezing temperature also decreases with salinity, with typical seawater freezing at around -1.9°C at atmospheric pressure.

The curvature of the isopycnals gives rise to a phenomenon known as cabbeling, or a contraction on mixing. Consider two water masses, A and B, with different Conservative Temperature and Absolute Salinity, which are plotted at different locations on the Θ-S_A diagram. Water mass A is fresher, but colder, than water mass B. However, near the surface when pressure is 0 dbar, the density of each water mass is less than or equal to 1027 kg/m³, as the points representing these water masses are on or left of the 27 kg/m³ isopycnal (red curves in above figure). The Conservative Temperature and Absolute Salinity of a mixture of two water masses will lie along the straight line (the 'mixing line') joining them in this diagram. However, the curvature of the isopycnals means that the density of this mixture may be to the right of the 27 kg/m³ isopycnal (i.e., its density may be greater than 1027 kg/m³). This increase in density is important in the Southern Ocean. There, different water masses on the surface can be brought together by ocean currents. The parcels mix, and after mixing together, the denser mixed water that results sinks below the surface.

If we draw this diagram at higher pressures (e.g., blue curves in figure), the general features remain the same, but quantitative aspects can change. An obvious change is that the freezing temperature decreases with pressure. Thus melt water from the base of thick Antarctic ice shelves, which is at the freezing point at depth when pressures are high, can become supercooled after rising upwards in the water column. Such waters are difficult to measure as instruments lowered into them rapidly become coated in ice crystals. Also, as pressure increases, the isopycnals do not remain in the same place in the diagram. Instead, densities increase with pressure and the isopycnal curves move leftwards. Isopycnals calculated at a pressure of 1100 dbar show that densities for a given Θ and S_A are about 5 kg/m³ higher than they are at the surface.

This increase in density with pressure is a problem when trying to decide which of two water parcels in the ocean, initially at different depths, would be lighter than the other if they were brought together. This is because the pressure effects often result in the largest apparent differences. Instead of comparing the in-situ densities, it is more useful to compare the densities as if both parcels are brought adiabatically to the same reference pressure. When this reference pressure is taken to be at the surface, the resulting measure of density is called potential density, denoted σ_θ.

However, in addition to generally moving leftwards as pressure increases, the isopycnals on this diagram also rotate slightly. The isopycnals calculated at 1100 dbar are tilted rightward relative to those calculated at 0 dbar. The density of warmer waters does not increase quite as much as the density of colder waters for the same change in pressure. Thus, two water parcels with different temperatures and salinities but the same density at the surface (i.e., with the same potential density) will in fact have slightly different densities at another depth even if their heat content and salinity remain the same. This is called the thermobaric effect. In deep convection areas, which are places in the Labrador and Greenland seas, and in some locations around Antarctica where surface water is cooled and eventually sinks to depths of several thousand meters, the thermobaric effect can be important in accelerating downward convection, resulting in large vertical displacements in the position of water parcels.

On the other hand, the rotation of isopycnals can also lead to mistakes in the interpretation of potential density. Consider water parcels A and B in above Figure again. Near the surface, parcel B is slightly denser than parcel A. However, at 1100 dbar, parcel A is slightly more dense than parcel B. If we are trying to decide which of the two water masses is heavier at depths near 1100 dbar, then comparing potential densities will give an incorrect answer. This difficulty is an important factor in correctly understanding water characteristics near the bottom of the South Atlantic. There, waters with larger potential density are observed to lie above waters with smaller potential density, which at first glance suggests a large-scale instability. However, the shallower water mass is heavier than the deeper water mass only if both are brought to the surface. If both are brought to the ocean bottom, the shallower water mass is lighter than the deeper water mass and one can conclude that the water column is stably stratified.

Chemical Oceanography

Chemical oceanography is concerned with the study of the dissolved elements in sea water and the ocean's numerous chemical and biochemical cycles. Topics of study include the origin and evolution of sea water, the origin of the sediment that covers the seafloor, the relationships between the myriad of chemical constituents of sea water, and the significance of changes in ocean chemistry (i.e., the influence of changing geology, including biological activity, and human-induced pollution).

Chemical oceanography can be further divided into focused areas of study. For example, marine chemistry is concerned with the composition of sea water. Marine geochemistry is additionally concerned with the chemistry of the precipitated rocks and sediment found on the ocean floor. Additionally, marine biogeochemistry is concerned with the role of organisms (particularly microorganisms) in the alteration or formation of geological features in the oceans.

The study of pollutants holds a high priority among many chemical oceanographers. Runoff of sewage, oil, fuel, and agricultural chemicals into the oceans decreases sea-water quality, particularly along the coast. At a local scale (i.e., beach or ocean/estuary interface), the decreasing water quality is more easily detected. In contrast, the global effect of ocean pollution is more difficult to determine. The full scope and the significance of the pollution-related changes are currently not clear, although chemical oceanographers are involved in clarifying the interaction between the ocean water with various pollutants, and with the ocean surface and the sea floor.

Another increasingly important aspect of chemical oceanography research concerns the study of the role of oceans in the global carbon cycle. The oceans are a major source and reservoir of carbon dioxide. Too much carbon dioxide in the atmosphere traps the escape of heat, leading to increasing global temperatures ("greenhouse effect"). The role of the oceans in potential global warming remains to be clarified.

Many elements are soluble in sea water; the oceans are major reservoirs of these elements. The study of trace elements (such as mercury and arsenic that are usually present in nature at very low levels) in sea water is important in the understanding of cycling of these elements between inorganic and organic processes. For example, naturally occurring elements such as mercury and arsenic are toxic to humans in high concentrations, and a deeper understanding of how the oceans contribute to potential human exposure (e.g., through consumption of mercury-laden fish) is gaining research importance.

Precise elemental studies may require sophisticated equipment and ultraclean sampling containers. Obtaining high-quality results can be time consuming and difficult, but the results have proved significant. Chemical oceanographers, for example, were among the scientists who first discovered and unraveled the unique ecosystem of hydrothermal vents that are present at the extremely cold, lightless bottom of the ocean floor.

A related area of chemical oceanography is concerned with the speciation of trace metals in ocean water. Some metals exist in a number of different forms, or species. Metals such as manganese, iron, nickel, and zinc form certain chemical species when organisms utilize the metals.

Biological Oceanography

Biological oceanography is a field of study that seeks to understand what controls the distribution and abundance of different types of marine life, and how living organisms influence and interact with processes in the oceans.

Biological oceanographers study all forms of life in the oceans, from microscopic plants and animals to fish and whales. In addition, biological oceanographers examine all forms of oceanic processes that involve living organisms. These include processes that occur at molecular scales, such as photosynthesis, respiration, and cycling of essential nutrients , to large scale processes such as effects of ocean currents on marine productivity.

A distinction is often made between the fields of biological oceanography and marine biology. Although there is considerable overlap between the two disciplines, the field of marine biology traditionally deals with the study of individual organisms, including their taxonomy, behavior, physiology and other aspects of their biology. In contrast, the emphasis of biological oceanography is the ocean and organisms as a system. As such, biological oceanographers tend to utilize a multidisciplinary approach, drawing on knowledge from various fields in addition to biology including, for example, physics, chemistry, and geology.

Tools and Technology

Biological oceanographers rely on a variety of tools and use a variety of approaches to aid them in their study of life in the sea. Some studies involve laboratory experiments with individual organisms. In other cases, the oceanographer must go into the water to directly sample and observe certain types of organisms such as zooplankton.

Other approaches involve underwater submersible vehicles to gain access to biological communities deep in the ocean, such as those associated with deep-sea hydrothermal vents. Many oceanographers use research vessels from which they lower instruments and specialized water sampling gear into the water. Biological oceanographers employ methods derived from various fields, including molecular biology, immunology, physiology, biochemistry, ecology, and many others.

In addition to making scientific observations, the biological oceanographer uses a variety of models to study the biology of the oceans. Theoretical models are used to examine problems in biological oceanography that cannot be answered through direct observation and measurement. Heuristic models are used to help to understand and explain an existing set of observations. Finally, some models are used to predict changes in biological processes that may occur because of natural and human-induced changes to the ocean environment.

Advances in technology have given biological oceanographers new insights about the living oceans. Lasers, fiber optics, high-speed digital video imaging and DNA microarrays are some of the high-tech "gadgets" that are used to study biological processes in the oceans. Robotic underwater vehicles reduce the risk and expense of manned submersibles while providing spectacular views of undersea communities. Other types of instruments are allowed to drift freely with ocean currents, towed behind a ship, or anchored at specific locations to provide detailed information over time and space. Among the most powerful tools available to biological oceanographers are satellite and airborne sensors, which provide large-scale views of the ocean and have greatly enriched the scientific understanding of biological processes and their relationship to physical phenomena.

Geological Oceanography

Geological oceanography is the study of Earth beneath the oceans. It involves geochemical, geophysical, sedimentological and paleontological investigations of the ocean floor and coastal margins. Geological oceanography includes exploring the ocean floor and the processes that form its canyons, valleys and mountains. Geological oceanographers research on the sea-floor spreading, plate tectonics, and oceanic circulation and climates. They examine the various ocean features such as rises and ridges, seamounts, trenches, etc. Geological oceanography is one of the broadest fields in the Earth Sciences and contains many sub-disciplines, including geophysics and plate tectonics, petrology and sedimentation processes, and micropaleontology and stratigraphy.

Sediment on the seafloor originates from a variety of sources, including biota from the overlying ocean water, eroded material from land transported to the ocean by rivers or wind, ash from volcanoes, and chemical precipitates derived directly from sea water. A very small amount of it even originates as interstellar dust. In short, the particles found in sediment on the seafloor vary considerably in composition and record a complex interplay of processes that have acted to form, transport, and preserve them.

Geological oceanographers have coined the terms "terrigenous" to describe those sediments derived from eroded material on land, "biogenic" for those derived from biological matter, "volcanogenic" for those that include significant amounts of ash, "hydrogenous" for those that precipitate directly from sea water and cosmogenic for those that comes from interstellar space.

The seafloor, however, is not a random arrangement of these different sediment types. Oceanographers have painstakingly mapped the distribution of sediment around the globe and have learned that at any given location the sediments provide important information regarding the history of the ocean as well as the overall state of climate on the Earth's surface. By studying how the heterogeneous composition of sediment varies as a function of geographic location and age, oceanographers are able to document the geologic and climatic conditions that are responsible for that sediment.

Oceanographers study sediment by taking long cylindrical cores, which individually can be as long as 18 to 30 meters (60 to 98 feet). Because the bottom of the ocean is extremely cold (only 1 to 3 degrees above freezing), the cores are stored in refrigerators onboard the research ship prior to being stored in large refrigerated repositories at shore-based laboratories.

Figure: A core sample of sediment from Chesapeake Bay can tell scientists about the oceanographic history of that particular location, including climate change, pollution, and past changes in erosion.

Regardless of which type of sediment, there are three processes that are responsible for its final composition: namely, the production of the sediment; its transport; and its preservation. It is important to differentiate between these three processes. For example, if a sedimentary particle is produced, but not preserved, there will be no resulting sedimentary record. Thus, only if material is produced and transported and preserved will marine sediment result.

The different combinations of each process' effectiveness result in a commensurate variety of sedimentation rates. Sediment can accumulate as slowly as 0.1 millimeter (0.04 inch) per 1,000 years (in the middle of the ocean where only wind-blown material is deposited) to as fast as 1 meter (3.25 feet) per year along continental margins . More typical deep-sea rates are on the order of several centimeters per 1,000 years.

Production of Sediment

The production of marine sediment is more complex than it may seem. Terrigenous sediment is produced by an interplay of chemical and physical weathering processes, which collectively serve to create small grains of material ranging in size from thousandths of millimeters to 1 or 2 millimeters(0.04 or 0.08 inch).

Physical weathering is caused by mechanical fracturing of rocks, such as that due to the freezing of water in cracks, and results in finer grained, compositionally similar examples of the original rock. On the other hand, chemical weathering, caused by the weak acid produced by the interaction of rainwater and atmospheric carbon dioxide, degrades the rock slowly and often produces fine-grained minerals that are compositionally distinct from the original rock.

Biogenic (biologically derived) sediment is produced by marine plankton, which are small, often microscopic, unicellular plants and animals that float in the surface waters of the ocean. The shells of these organisms are made of either calcium carbonate ($CaCO_3$) or silica (SiO_2). Although ubiquitous, particularly elevated concentrations of such organisms are most commonly found in biologically productive waters such as the Equatorial Pacific, or the Southern Ocean ringing the continent of Antarctica.

Volcanic ash is produced during volcanic eruptions, as can be seen in the billowing ejected material from many volcanoes. Cosmogenic material is the remains of primordial material left over from the creation of the solar system (and perhaps from beyond) and, although very low in abundance, is ubiquitously distributed.

The production of hydrogenous sediment is most difficult to visualize, but involves either the slow precipitation of dissolved chemicals from sea water or the leaching of chemical elements from rocks that have extremely hot sea water (greater than 300°C [572°F]) circulating through them along mid-ocean ridges. When these hot solutions are injected into the cold sea water the leached chemical elements precipitate from the cooling water, leading to hydrothermal sediments near the mid-ocean ridge that are enriched in iron, manganese, copper, zinc, and other metals.

Transport of Sediment

The transport of sediment depends on its grain size and the original location where it was produced. Terrigenous sediment can be transported to the deep sea via rivers or by wind. Material transported by rivers most commonly ends up deposited on the continental margin, the shallow portions of the ocean that are within several hundred kilometers of land. When continental margin deposits accumulate fast and get overly steep, or when an earthquake or storm causes the sediment to be re-suspended, turbidity currents provide additional transport out to the deep sea. The re-suspension of the sediment into the bottom water causes it to be denser than the overlying water, and thus these turbidity currents flow downslope to the more distant ocean basin.

The transport of sediment by wind is also extremely significant, and is particularly relevant to studies of Earth's climate in the past. When the Earth's climate is relatively dry (arid), such as during glacial periods, the land surface tends to be more dusty, and thus during such periods there will be more windblown terrigenous material delivered to the deep ocean. Also, during such time periods the wind speed tends to be higher, and thus terrigenous grains that are slightly larger than usual are preferentially transported. Thus, by examining the amount of dust, as well as its grain size, in the different layers of a sediment core, oceanographers learn how arid the land surface was at a given time, as well as how fast the average wind speeds were.

Although such dust is essentially invisible to the human eye, its transport is still an important and long-ranging process. For example, dust derived from the Sahara in North Africa is easily observed in Miami, Florida and even in the eastern Pacific Ocean. Moreover, volcanic ash ejected tens of kilometers into the atmosphere during the largest eruptions can be transported by winds all around the globe.

The microscopic shells of the plankton do not just simply fall to the seafloor. In fact, because they are so small, the plankton may not be able to fall individually. Oceanographers learn how such sediment is delivered to the seafloor by suspending sediment traps in the ocean. These traps are essentially large funnels, up to 1 or 2 meters (3.3 to 6.6 feet) in diameter, that collect the material as it falls through sea water.

By examining the material trapped by these instruments, it was discovered that plankton shells are delivered to seafloor by "bio-packaging" via fecal pellets. In other words, various microorganisms that eat other plankton excrete their shells in fecal pellets. These "biopackaged" fecal pellets are large enough (0.2-1.5 millimeters, or 0.008-0.059 inch) and dense enough to sink to the seafloor, where they become part of the sediment.

Figure: Wind-swept desert sands not only produce a cooling effect due to deflection of incoming solar radiation, but they also deposit sand, silt and dust on the ocean surface and ultimately on the ocean floor. Through mineralogical and chemical analysis, scientists can recreate historical patterns in climate and geological development.

Hydrothermal sediment is largely localized to within less than 10 kilometers (6.2 miles) of the mid-ocean ridge. The concentration of metals in these sediments decreases with distance from a ridge, yet small amounts can be found up to 500 to 1000 kilometers (300-600 miles) away.

Preservation of Sediment

Terrigenous sediment, whether it be delivered by rivers or wind, is not altered significantly on the seafloor and thus is well-preserved. During very deep burial (e.g., 5 kilometers, or 3 miles, below the seafloor), the terrigenous grains can be altered into different minerals, but this does not occur while the grains are lying on the seafloor and is generally a more important process for geologists rather than oceanographers.

Biogenic sediment, on the other hand, is very poorly preserved on the seafloor. The degradation of biogenic sediments is a complex, largely chemical suite of processes. For example, significantly less than 1 percent of the siliceous plankton that is biopackaged to the seafloor is preserved. This is because sea water is undersaturated with respect to silica. Therefore, the siliceous planktons are living in an environment that is corrosive to their shells.

While the plankton is alive, the shell is surrounded by organic protoplasm that protects it from the corrosive sea water. After death, however, even if bio-packaged, this organic coating will degrade, exposing the shell of the siliceous plankton. When exposed to sea water, the shell will dissolve.

This process occurs over all depth and temperature ranges throughout the global ocean. Thus, the only region of the seafloor where biogenic silica appreciably accumulates is where the production of biogenic silica is so enormous that it overwhelms the amount that is dissolved. In the modern oceans, this occurs at high latitudes in the North Pacific and Southern Ocean and the Equatorial Pacific Ocean.

Plankton with shells made of calcium carbonate also commonly dissolves, but not as commonly as siliceous plankton. The dissolution of carbonate plankton is controlled by water depth and water temperature. Water depth and hydrostatic pressure correlate with each other—at greater depths there is greater pressure. At greater pressures, the solubility of carbon dioxide gas increases. An excellent analogy of this process is observed in a bottled carbonated beverage that is under pressure until opened—when the pressure is released the carbon dioxide comes out of solution and bubbles form. Similarly, at the great depths of the deepest seafloor, the solubility of carbon dioxide increases so much that calcium carbonate sediment may dissolve. This dissolution is also facilitated at the lower temperatures of the deep sea.

The converse is also true. At shallow water depths (that is, lower pressure) the carbonate does not dissolve and the warmer water temperatures (along with the increased light for photosynthesis) each serve to enhance the construction and preservation of coral reefs and other carbonate-producing biota. Thus, there is both a depth and latitudinal effect on the distribution of carbonate sediments due to their influence on temperature and pressure.

Paleoceanography

Paleoceanography is the scientific study of Earth's oceanographic history involving the analysis of the ocean's sedimentary record, the history of tectonic plate motions, glacial changes, and established relationships between present sedimentation patterns and environmental factors.

Prior to the breakup of Pangea, one enormous ocean, Panthalassa, existed on Earth. Currents in this ocean would have been simple and slow, and Earth's climate was, in all likelihood, warmer than today. The Tethys seaway formed as Pangea broke into Gondwana and Laurasia. In the narrow ocean basins of the central North Atlantic, restricted ocean circulation favoured deposition of evaporites (halite, gypsum, anhydrite, and other less abundant salts). Evaporites also were deposited some 100 million years ago in the equatorial regions of the South Atlantic during the early opening of this ocean.

Sequences of organic-rich, black shales were deposited during the early phases of spreading in the North and South Atlantic. These sediments indicate anoxic conditions in the deep ocean waters. The oceans must have been well stratified into dense layers to prevent the overturning and mixing required replacing depleted oxygen. Black shales also were deposited in the older areas of the eastern Indian Ocean.

During the time interval between 200 and 65 million years ago, but especially from about 100 to 65 million years ago, microplankton abundance and diversity increased enormously in the oceans. This resulted in increased deposition of biogenic sediments in the ocean basin. During the Cretaceous Period (145.5 to 65.5 million years ago), sea level was often high, and shallow seas lapped onto the continents. This may have provided an environment favourable to the explosion in the numbers of species of foraminiferans, diatoms, and calcareous nannoplankton (single-celled, photosynthetic organisms with shells made up of calcium carbonate plates called coccoliths). Increased abundance of calcareous nannoplankton shifted the locus of carbonate sedimentation from shallow seas to the deep ocean. The end of the Cretaceous Period is marked by a sudden extinction of many life-forms on Earth, and marine organisms were no exception. Coccolithophores (calcareous nannoplankton) and planktonic foraminiferans were particularly affected, and only a few species survived. Ocean sediments were suddenly less biogenic, and clays became widespread.

After the Cretaceous Period, Earth underwent a gradual cooling, especially at high latitudes. Deep-sea sedimentation changed as thermohaline bottom-water circulation became fully developed. The Calcite Compensation Depth (CCD), the level at which the rate of carbonate accumulation equals the rate of carbonate dissolution, rose in the Pacific and dropped in the Atlantic as a result of changes in thermohaline circulation. An event of major significance was the spreading away of Australia from Antarctica beginning about 58 million years ago. This separation initiated limited circum-Antarctic circulation, which isolated Antarctica from the warmer oceans to the north, and led to cooling, which set the stage for later major glaciation.

At the boundary between the Eocene and Oligocene epochs (33.9 million years ago), Antarctic Bottom Water (AABW) began to form, resulting in greatly decreased bottom-water temperatures in both the Pacific and Atlantic oceans. Bottom-living organisms were strongly affected, and the CCD suddenly dropped from about 3,500 metres (about 11,500 feet) to approximately 4,000 to 5,000 metres (13,000 to 16,000 feet) in the Pacific. Bottom-water temperatures were generally warm, 12 to 15 °C (54 to 59 °F), during the time preceding this event. In a study of deep-sea sediment core material from near Antarctica, New Zealand Earth scientist J.P. Kennett and American oceanographer Lowell D. Stott discovered that there was a period between roughly 50 and 35 million years ago when deep waters were very warm (20 °C [68 °F]) and salty. The origin of these ocean waters was most likely in the low latitudes and resulted from high evaporation rates there.

The modern oceans are distinguished by very cold bottom water. The gradual changes toward this condition began 10 million years after the origination of AABW. Particularly significant among these changes was the closing of the Tethys seaway as Australia and several microcontinents moved north into the Indonesian region. Also, Australia moved far enough North that circum-Antarctic surface circulation became fully established.

The modern ocean circulation patterns and basin shapes were mostly in place by the beginning of the Miocene Epoch (about 23 million years ago). An exception was an ocean connection between the Pacific and Caribbean Sea in Central America that persisted until about three million years ago. Major and probably permanent ice sheets on Antarctica formed during the Miocene Epoch, and glacial sediments began to dominate the seafloor surrounding the continent shortly thereafter. Siliceous oozes also became widespread around Antarctica. Siliceous sedimentation increased in this area at the expense of siliceous sedimentation in equatorial regions. Ocean circulation became more vigorous, global climate became cooler, and sedimentation rates in the ocean basins increased. Planktonic microorganisms were segregated into latitudinal belts. Bottom-water flow north through the Drake Passage between South America and Antarctica began in the Miocene Epoch, resulting in erosion and nondeposition of sediments in the southwest Atlantic and southeast Pacific oceans. Also during the Miocene Epoch rifting between Greenland and Europe had progressed to a point where a connection was established between the North Atlantic and the Norwegian Sea. This resulted in the formation of North Atlantic Deep Water, which began flowing south along the continental rise of North America at this time. Sediments redistributed and deposited by this deep current are called contourites and have been extensively studied by American geologists Bruce Heezen, Charles D. Hollister, and Brian E. Tucholke, among others.

Sudden global cooling set in near the end of the Miocene Epoch some six million years ago. The strength of ocean circulation must have increased, as evidence of increased upwelling and biological productivity is present in ocean sediments. Diatomaceous sediments were deposited in abundance around the rim of the Pacific. This cooling event is synchronous with a drop in sea level, thought to be about 40 or 50 metres (130 to 165 feet) by various authorities, and probably corresponds to the further growth of the Antarctic ice sheet. This lowered sea level, coupled with the closure of narrow seaways probably due to plate movements, isolated the Mediterranean Sea. Subsequently, the sea dried up, leaving evaporite deposits on its floor. The Swiss geologist Kenneth J. Hsü and the American oceanographer William B.F. Ryan have concluded that the Mediterranean probably dried up about 40 times as seaways opened and closed between six and five million years ago. This evaporation removed about 6 percent of the salt from the world ocean, which raised the freezing point of seawater and promoted further growth of the sea ice surrounding Antarctica.

Enormous ice sheets emerged in the Northern Hemisphere between three and two million years ago, and the succession of Quaternary glaciations began at 1.6 million years ago. The exact cause of the glacial period is unclear, but it is most likely related to the variability in solar isolation, increased mountain building, and an intensification of the Gulf Stream at three million years ago due to the closing off of the Pacific-Caribbean ocean connection in Central America. The Quaternary glaciations, of which there were probably 30 episodes, left the most dramatic record in ocean sediments of any event in the previous 200 million years. Terrigenous sedimentation rates greatly increased in response to fluctuations in sea level of up to 100 metres (about 300 feet) and a more extreme climate. Biogenic sedimentation also increased and fluctuated with the glacial episodes. Deep-sea erosion began in many places as a result of intensified bottom-water circulation.

Tools and Methods of Oceanography

Argo (Oceanography)

Argo is broad-scale global array of temperature/salinity profiling floats, is a major component of the ocean observing system. Deployment of floats began in 2000. Conceptually, Argo builds on the existing upper-ocean thermal networks, extending their spatial and temporal coverage, depth range and accuracy, and enhancing them through addition of salinity and velocity measurements. The name Argo is chosen to emphasize the strong complementary relationship of the global float array with the Jason altimeter mission. For the first time, the physical state of the upper ocean is systematically measured and assimilated in near real-time.

The main objective of the Argo program is to provide a quantitative description of the evolving state of the upper 2000m of the ocean by collecting profiles of temperature and salinity from the surface to 2000m. Argo data is used for initialization of ocean and coupled (ocean atmosphere) forecast models, data assimilation and dynamical model testing. It has been demonstrated that the assimilation of Argo data in the models improved weather and climate forecast. Profiling float data have an enormous range of applications. The more distinctive categories for Argo data use are: educational uses, operational uses and research uses.

An Argo Float is a steel float weighing about 25kg, with instruments inside. An Argo float is able to do three things:

An Argo Float:

- Is able to change its buoyancy so it floats at different depths.

- Is able to take measurements of sea water.

- Is able to send information to a satellite.

Argo Float Buoyancy

By changing its own buoyancy, an Argo Float is able to float at different depths. It changes its floating depth regularly every 10 days like this:

	Where is the Argo Float?	What is the Argo Float doing?	Approximate time
Day 1	On surface	Floating	6 hours
	Going down	Descending	6 hours
Day 2-9	At 1,000m	Floating	9 days
	Going down	Descending	2 hours
	At 2,000m	Floating	2 hours
Day 10	Going up	Ascending	8 hours
	On surface (same as day 1)		

Changes in technology now allow the Argo Float to send data to satellites in 20 minutes rather than several hours. This means it has to spend less time on the surface where bad things can happen to it e.g. getting hit by other floating objects.

This 10 day up-and-down is called a 'cycle'. At the end of every cycle the Argo Float sends its information to satellites. This information is called a 'profile'.

Argo Measurements

While the Argo Float is underwater it measures three things about the ocean:

An Argo Float measures:

- Salinity (amount of salt in the water) measures in parts per thousand.
- Temperature in degrees Centigrade (°C).
- Pressure in decibars (db).

1decibar (db) is equal to 1 metre of depth. So 1,000db is about 1,000m below the sea surface.

Argo Information to Satellites

While the Argo Float is on the sea surface it sends its information to a satellite. This information is called 'data'. When the Argo Float comes to the surface every 10 days it sends the following data to a satellite:

- Temperature and depth,
- Salinity and depth,
- Position.

The data uploaded to satellites is collected while the Argo Float travels from 2,000m underwater to the surface. The data describes a 'profile' of the ocean. That means it shows the changing temperature and salinity as the Argo Float rises to the surface. Over a million Argo Float profiles have been uploaded since the first Argo Float was deployed in the year 2000. Have a look at the image on this page of a typical Argo Float profile. The position of the Argo float tells us how far it has moved since the last time it came to the sea surface. The position is given as latitude and longitude.

Regular Argo Floats

Regular Argo Floats are the ones described above. They are steel cylinders about 1.2m long and weighing about 25kg. They are designed to be strong enough to descend to 2,000m below the surface of the sea. At a depth below 2,600m they will crush and be destroyed.

Deep Argo Float

New Argo profilers can reach depths of 4000 and 6000 m well under the 2000 m of the initial models of the core Argo mission. An operational depth of 2000 dbar gives access to 51% of the seawater volume, while 4000 and 6000 dbar represents 88% and 98%, respectively. The choice of the operational depth has implied significantly different developments and technology options: for an operational depth of 4000 dbar, similar solutions to those used for the 2000 dbar design can be used after some innovation or adaptation. The performance can be estimated and the nominal number of 150 Argo cycles (without dissolved oxygen sensor) can be reached for the "Deep ARVOR." For an operational depth of 6000 dbar, glass sphere flotation devices, such as those manufactured by Benthos (Teledyne group), are used today. These housings are different from those of the "ordinary" profiling Argo floats. Glass spheres have a greater stiffness than the tubular solution. The cycle of profiles can be one dive down to 4000 or 6000 m every "N" dives down to 2000 m: this solution allows more profiles to be performed, as required by the initial Argo recommendations.

Table: Deep Argo Floats Commercial Offer in 2017.

4000 m/TS SBE 41 CP and dissolved oxygen/150 cycles from 4000 m (CTD only)∗ 30 kg dry weight.	4000 m/TS SBE 41 CP/75 mixed cycles 2000–4000∗ Approx. 50 kg dry weight.
Developed in industrial partnership with Ifremer/nke Instrumentation.	Developed jointly by Jamstec and Tsurumi-Seiki Co. Ltd.
6000 m/dry weight is approximately 51 kg Apex Deep will perform 150 profiles ∗∗from 6000 m∗∗ with the SBE 61 CTD in continuous/Source Webb research.	6000 m/TS SBE 61./dry weight is 27 kg/ 190 cycles from 6000 m∗∗ Developed by Scripps Institute of Oceanography and manufactured by MRVSYS.

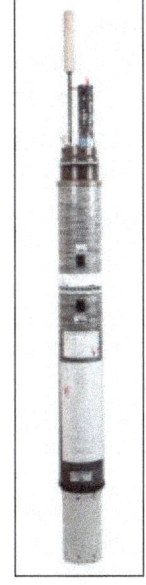

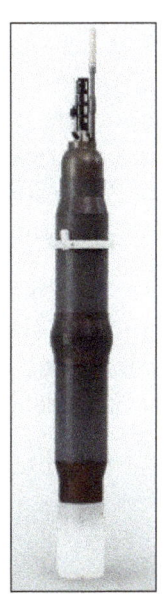

Figure: Deep Arvor.   Figure: Deep Ninja.

Figure: Apex Deep.   Figure: Deep Solo.

Today, these deep-profiling floats are equipped to provide measurements of PTS and dissolved oxygen, but other measurements will soon be implemented for studies of great depths to satisfy the scientific demand. The deep-profiling floats will probably also be used in the industrial sector and deep oil and gas exploitation as autonomous stations for environmental monitoring during the exploitation phase. By the end of 2017, only a few dozen deep Argo floats were deployed and full maturity of these instruments has not yet been achieved.

Component of Argo Float

The different models of Argo floats made around the world have differences in both hardware and software, but mostly operate in a similar fashion.

- Antenna: All Argo floats are topped with an antenna that allows them to communicate with satellites to send their data, obtain a position and find out if there is a new mission. Over half of the current Argo fleet use GPS to get their position, and Iridium satellites to transfer the data collected in the last cycle and download new mission instructions if available. Using GPS and Iridium allows floats to spend less time transmitting data at the surface, which reduces the risk of washing up on shore. This advance means that the Argo Program can put floats into marginal seas such as the Mediterranean Sea and the Gulf of Mexico. Using Iridium has also allowed for more data to be sent back, including more observations and more information about the health of the float. Instead of Iridium and GPS, some floats use System Argos to both obtain a position and send back data. Due to data transmission rates and location, these floats usually need to spend 8 – 12 hours on the surface to guarantee a successful data transmission and a position accurate to ~100m.

- Conductivity, Temperature, Depth Sensor (CTD): On the top of every Argo float is a CTD which measures temperature within an accuracy of 0.001 degree C , pressure (closely related to depth) within 0.1 dbar, and calculate salinity using conductivity, temperature, and pressure within 0.001 psu (practical salinity units).

- Controller: Each Argo float is controlled by a small computer that contains a program to run the float.

- Internal Reservoir: This area of the float stores oil, when it is not inflating the external bladder.

- Hydraulic Pump: This mechanism moves oil between the internal reservoir and the external bladder to control the buoyancy of the float.

- Batteries: Modern Argo floats use lithium batteries to power the pumps, sensors, controller, and communication system. The battery power available is the main limitation on a float's operational lifetime.

- External Bladder: Roughly the size of a large grapefruit, this bladder is the mechanism by which Argo floats control their buoyancy or depth in the ocean. Inflating the bladder with oil from the internal reservoir increases the volume of the float without changing the mass, thereby lowering density and causing the float to rise. Pumping oil out of the bladder back into the reservoir will increase the float's density, allowing it to sink to a specified depth.

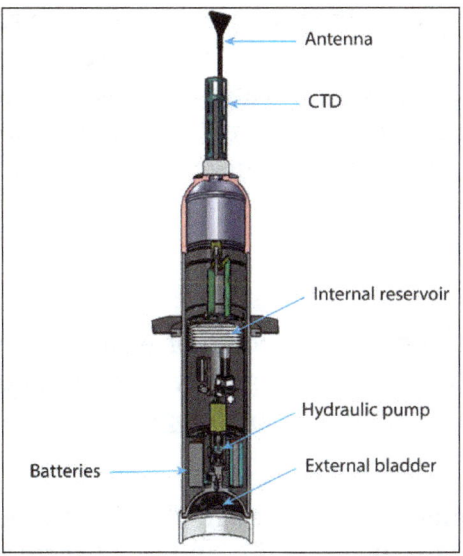

Echo Sounding

The most common system for measuring water depth, and preventing collisions with unseen underwater rocks, reefs, etc., is the echosounder. These sonar systems use a transducer that is usually mounted on the bottom of a ship. Sound pulses are sent from the transducer straight down into the water. The sound reflects off the seafloor and returns to the transducer. The time the sound takes to travel to the bottom and back is used to calculate the distance to the seafloor. Water depth is estimated by using the speed of sound through the water (approximately 1,500 meters per second) and a simple calculation:

$$\text{Distance} = \text{speed} \times \text{time}/2$$

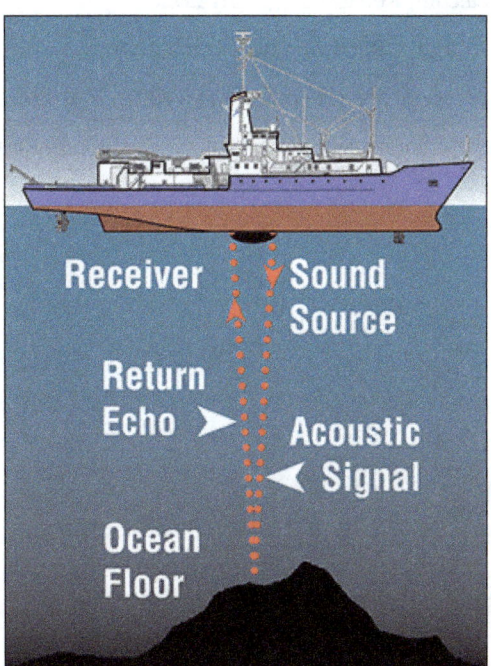

Figure: Echosounder calculate water depth by measuring the time it takes for the acoustic signal to reach the bottom and the echo to return to the ship.

The product is divided by two because the measured time is the round-trip time (from the transducer to the seafloor and back to the transducer). The faster the sound pulses return to the transducer from the ocean floor, the shallower the water depth is and the higher the elevation of the sea floor. The sound pulses are sent out regularly as the ship moves along the surface, which produces a line showing the depth of the ocean beneath the ship. This continuous depth data is used to create bathymetry maps of the survey area.

Bathymetry

Bathymetry is the study of the "beds" or "floors" of water bodies, including the ocean, rivers, streams, and lakes. The term "bathymetry" originally referred to the ocean's depth relative to sea level, although it has come to mean "submarine topography," or the depths and shapes of underwater terrain.

In the same way that topographic maps represent the three-dimensional features (or relief) of overland terrain, bathymetric maps illustrate the land that lies underwater. Variations in sea-floor relief may be depicted by color and contour lines called depth contours or isobaths.

Bathymetry is the foundation of the science of hydrography, which measures the physical features of a water body. Hydrography includes not only bathymetry, but also the shape and features of the shoreline; the characteristics of tides, currents, and waves; and the physical and chemical properties of the water itself.

References

- Key-physical-variables-in-the-ocean-temperature-102805293: nature.com, Retrieved 15, January 2020

- Paleoceanography: britannica.com, Retrieved 23, July 2020

- What-is-an-argo-float: learnz.org.nz, Retrieved 06, March 2020

- Earth-and-planetary-sciences: sciencedirect.com, Retrieved 18, May 2020

- Echosounder, observing-the-sea-floor, technology-gallery: dosits.org, Retrieved 04, February 2020

Chapter 2

Biological Oceanography

Biological oceanography refers to the study of ocean ecology which is focused on the interaction of organisms with various aspects of the oceanographic system. It includes the study of marine ecosystem, and marine flora and fauna. This chapter closely examines the key concepts of biological oceanography to provide an extensive understanding of the subject.

The basic ecological concepts are central to many studies of biological oceanography. The study of marine life, habitat, interactions, abiotic environment, phytoplankton and primary production, zooplankton, migrations and changes, energy flow & mineral cycling, marine food chains, food webs, nektons, marine reptiles, mammals, seabirds, mariculture, Benthic plants and animals, inter-tidal environments, beaches, coral reefs, estuaries and mangroves are all studied under biological oceanography. Deep sea ecology and marine pollution are also the other two major important areas of study under biological oceanography.

Marine Ecosystem

Marine ecosystems occupy specific areas within the ocean. Some are productive near shore regions, including estuaries, salt marshes, and mangrove forests. Others appear to be completely barren, like the ocean floor. The hydrosphere connects all freshwater and saltwater systems. The high salt content (salinity) and global circulation make marine ecosystems different from other aquatic ecosystems. Other physical factors that determine the distribution of marine ecosystems are geology, temperature, tides, light availability, and geography. Marine ecosystems include: abyssal plain (deep sea coral, whale fall, and brine pool), Antarctic, Arctic, coral reef, deep sea (abyssal water column), hydrothermal vent, kelp forest, mangrove, Open Ocean, rocky shore, salt marsh and mudflat, and sandy shore. Some of these regions are very productive. Others are in constant darkness where photosynthesis cannot occur. Some marine ecosystems go through extreme changes in temperature, light availability, oxygen levels, and other factors on a daily basis. Others are fairly stable and only change slightly at different seasons. The organisms that inhabit marine ecosystems are as diverse as the ecosystems themselves. They must be highly adapted to the physical conditions of the ecosystems in which they live.

The Abyssal Plain ecosystem is one of the least explored regions of the Earth. The abyssal plain runs along the ocean floor between the base of a continental rise and a mid-ocean ridge. The abyssal plain is a harsh environment located at depths between 3,000 and 6,000 meters (10,000–20,000 feet). Little to no light reaches the abyssal plain. High pressures, extreme cold, low oxygen levels, and scarce food availability also characterize this ecosystem. Despite these harsh conditions, research has shown that the abyssal plain can include several different communities of organisms. These include deep sea corals, whale falls, and brine pools.

Deep (cold) water coral communities are distributed globally from coastal Antarctica to the Arctic Circle. They are in the bathypelagic zone and not typically found below 2,000 meters (6,600 feet). They are made up of hard corals, black and horny corals, and soft corals, including gorgonians. Like warm tropical reefs, the hard corals in deep water develop a reef structure that supports a highly diverse community of fish, crabs, urchins, and worms. Unlike warm reefs, however, they lack symbiotic algae (zooxanthellae) because they are in such deep, cold, and often dark water. This is why deep water corals are typically suspension feeders, capturing plankton and organic matter (detritus) out of the water column.

A whale fall community is created when a dead whale sinks and transports nutrients to the sea floor. Whale falls can exist below any waters where whales are found, especially along their migratory routes and feeding grounds. These regions tend to be along continental shelves and the deep ocean. Whale fall communities are found in the abyssopelagic zone. This zone is usually sparsely populated and minimally productive, but whale fall communities supply organic matter (up to 160 tons from one whale) and lead to the development of complex animal communities and feeding relationships. In this way, they are similar to hydrothermal vent and seep communities. After a whale falls, the first feeders (hagfish, certain crabs) are the scavengers that eat the soft tissue. Then worms, snails, and other mollusks feed on the bones and organic matter in the surrounding sediment. As bacteria increase and decompose the whale, sulfur chemicals are released and used by chemosynthetic bacteria to make organic matter. That organic matter can then be consumed for energy by other organisms. This succession of feeders and whale fall community development can last for several years.

Brine pools are areas on the seafloor (abyssopelagic zone) that have distinct surfaces and shorelines. They are like small lakes. They can form within the surrounding ocean water because of extremely high salinities that make the brine pool water much denser than the surrounding abyssal ocean water. The more dense water sinks, forming pools on the surface of the abyssal plain. The brine pool contains high levels of methane gas that are used as chemical energy for surrounding mussel beds. The mussel beds then support communities of fish, crustaceans, and bacteria.

The Antarctic ecosystem of the Southern Ocean was once thought to be rather barren due to ice cover and extremely cold temperatures. Sunlight exposure in the Antarctic ranges from zero to 24 hours in a single day. These conditions make the Antarctic a harsh environment. However, research has shown that Antarctic waters are highly productive and support a diverse array of marine organisms. The Antarctic food web is the simplest on Earth and is based on one key species—krill. Krill are planktonic shrimp-like creatures. The Southern Ocean contains an estimated 500 million tons of krill. Krill is the most abundant animal in the world. Antarctica's whales, seals, penguins, albatross, petrels, fish, and squid depend on the presence of krill to support the food web. In turn, krill feed on phytoplankton and algae mats under the ice. Colder winters mean more pack ice and more algae mats. This is good for the krill when less phytoplankton is available. Without enough algae to eat, krill can die off and create a chain reaction throughout the entire food web. Researchers worry that global warming and decreasing pack ice threatens a fragile Antarctic ecosystem that is so dependent upon a single species.

The Arctic ecosystem of the northern polar region is characterized by extreme seasonality of light and year-round ice cover. Diverse arrays of highly adapted organisms interact and thrive there. Arctic Ocean species are distributed throughout the water column, within the ice, and along the seabed. Worms, copepods, bacteria, and phytoplankton live within pores and channels of the ice and at the ice-water interface. They form the base of the Arctic food web. The Arctic seafloor is comprised of numerous invertebrates, including sea stars, worms, sponges, mollusks, crabs, and anemones. In shallower depths where sunlight is available, macrophytes (large algae) are also found. Phytoplankton, copepods and other zooplankton, jellies, and fish are found in the water column between the Arctic seafloor and ice. These organisms support a number of Arctic predators, including birds, foxes, seals, and polar bears. The Arctic ecosystem contains highly adapted species and a complex food web. It is a fragile ecosystem that is vulnerable to threats from climate change, decreased ice cover, and human impacts.

Coral Reefs primarily occur throughout the warm tropical and subtropical regions of the ocean. These areas are at latitudes between the Tropic of Capricorn and Tropic of Cancer. Some corals can live in colder, deeper waters. Corals can be classified as hard (stony) or soft, and are made up of tiny animals, called polyps. Some corals contain a single polyp but most are colonial. Colonial corals contain thousands of polyps. Examples of flexible soft corals include the sea fans, sea whips, and sea plumes. True soft corals include black, mushroom, and tree corals. Stony, or reef-building, corals produce calcium carbonate skeletons and are responsible for creating the structure of the reef. Common types of stony corals include brain corals, pillar corals, plate corals, and branching corals like elkhorn and staghorn. Reef-building coral polyps divide as they grow, forming layer upon layer. Symbiotic algae (zooxanthellae) live within the tissues of reef-building corals. Some coral reefs are thousands of years old. Coral reefs are one of the most biologically diverse ecosystems in the world and are home to everything from sponges and jellies to octopus, manta rays, and sharks. They also provide spawning, nursery, refuge, and feeding areas for many marine organisms. The world's reefs contain over 4,000 different fish species and hundreds of coral species. Corals are fragile systems and their health or growth can be impacted by several factors. These factors include increased temperatures and nutrient runoff, decreased water clarity, and wave action from storms.

The Deep Sea (abyssal water column) ecosystem covers the vast zone of Open Ocean that is below 200 meters (656 feet). It includes the abyssal zone between 4,000 and 6,000 meters (13,000–20,000 feet). This ecosystem contains 98 percent of all the living space on Earth. The deep sea is a harsh environment. It is characterized by little-to-no light penetration, high pressures, extreme cold, low oxygen levels, and scarce food availability. The deep sea is less productive and populated than other marine ecosystems. The lack of light for photosynthesis results in a food web that depends on dead organisms and organic matter (detritus) sinking down from upper layers of the ocean. Several organisms are well-adapted to survive in the deep sea. In order to "see" in the dark and find food and mates, many organisms use bioluminescence (comb jellies, squid, anglerfish) or have large eyes (squid, octopus) and a strong sense of smell (fish). Many fish and invertebrates hide from predators by being black, silver, red, or translucent in color (siphonophores, jellies, mysid shrimp). Organisms that are silver or bioluminescent use their coloration to frighten or confuse predators. Their coloration also helps them communicate with species of their own kind. Many organisms have a soft, flexible body structure (siphonophore, dumbo octopus) so they can handle the high pressures of the deep sea. It is difficult to access and explore the deep sea. However, scientists believe that there are many more deep sea organisms left to be discovered.

Hydrothermal Vents are fissures (openings) in the ocean floor which release hot, mineral rich water. This vent water can reach temperatures of up to 750° F. There are two types of hydrothermal vents. The hottest of the vents, called "black smokers," spew a dark "smoke" composed mostly of iron and sulfide. The "white smokers" release a cooler, lighter material composed of compounds including barium, calcium, and silicon. Hydrothermal vents are found in both the Pacific and Atlantic Oceans at an average underwater depth of about 2,100 meters (7,000 feet). They are concentrated along the Mid-Ocean Ridge. The Mid-Ocean Ridge is the underwater mountain chain that winds its way around the globe. Hydrothermal vent ecosystems support familiar, yet unique and specifically-adapted life forms. Vent organisms include several species of bacteria, tubeworms, giant mussels, crabs, lobster, octopus, and even fish. Hundreds of species of animals have been identified in the hydrothermal vent ecosystems around the world. No light is available to support photosynthesis by marine algae or plants, but primary productivity occurs through chemosynthesis. This process occurs when bacteria-like organisms (archaea) turn chemical energy from the vents into useable energy. This process drives the entire hydrothermal vent food web.

Kelp Forests are coastal habitats found worldwide. They are most abundant along the rocky eastern coasts of the Pacific Ocean, from Baja California to Alaska, at depths of 2 to 27 meters (6 to 90 feet). They are often called underwater "forests" because they are layered and dominated by two canopy-forming kelp species (giant kelp and bull kelp). They also contain several understory kelp species. Kelp species are seaweeds or brown macro-algae (large algae) that prefer clear and cold, nutrient-rich waters. They can grow as much as 50 centimeters (20 inches) per day. They are one of the most productive and dynamic marine ecosystems. They provide a protected and biomass-rich habitat that supports a complex food web. Kelp forest food webs include several fish species, numerous invertebrates (snails, crabs, lobster, sea stars, sea urchins, worms), and large mammals (sea lions). Kelp forests have high levels of productivity and biomass. They are vulnerable to wave action, increased turbidity, and changes in salinity and temperature.

Mangrove ecosystems are comprised of salt-tolerant, woody mangrove trees and shrubs. They are located in shallow, low-oxygen sandy or muddy areas along shorelines. There are over 80 different species of mangrove trees throughout the tropical and subtropical zones of North and South America, Africa, the Middle East, Asia, and Oceania. The U.S. only has three species: red, black, and white mangroves. Black and white mangroves are less salt-tolerant than the red mangrove and are found farther from the water's edge. They have special adaptations to help them obtain oxygen and release excess salt. Red mangroves are the most salt-tolerant and are found closest to the water's edge. They are often submerged in shallow water with a thick, partially exposed network of roots (prop roots) that grow down from their branches. Their roots serve many important functions. They stabilize the shoreline by absorbing wave action and decreasing water flow. This allows sediments to accumulate. This prevents excess sediment and nutrients from reaching nearby seagrass and coral reef ecosystems. The prop roots also serve as a substrate (place of attachment) for numerous species of sponges, tunicates, algae, and shellfish. The mangrove forest provides a complex habitat for many ecologically and economically important organisms, including algae, sponges, barnacles, snails, mussels, crabs, fish, and birds. Many of these organisms use the mangrove for protection and feeding, and also as a nursery ground. They will then migrate to other marine ecosystems (coral reefs, Open Ocean, sandy shores) as adults. This makes mangrove ecosystems a vital part of maintaining fisheries within and adjacent to mangroves. Mangroves once covered vast areas of coastlines, but coastal development and pollution have destroyed significant amounts of mangrove habitat throughout the world.

Open Ocean is the largest marine ecosystem. It contains approximately 65 percent of the volume of the world ocean. It extends from the edge of the continental shelf outward and encompasses the entire water column. The open ocean zone generally refers to the upper 200 meters (656 feet) of water. This distinguishes it from the deep sea ecosystem below. It is a highly diverse and dynamic ecosystem that contains a wide variety of life. The diversity of open ocean organisms ranges from mega fauna, or large animals like sharks, whales, dolphin, and sea turtles to microscopic plankton and small schooling fish. Sea birds and large migratory fish also play an important part in this ecosystem. Although the mega fauna are large and seem to dominate, invertebrate species actually make up over 95 percent of the animal species found in the open ocean. Large populations of plankton drift along on ocean currents and form the base of the open ocean food web. Open ocean currents carry nutrients to different parts of the ocean. Currents also help some animals migrate and allow others to distribute their eggs throughout the ocean. Even though the abiotic conditions (light, temperature, salinity, circulation) of the open ocean are fairly consistent, there are areas where life is both abundant and sparse.

Rocky Shore ecosystems are coastal areas with solid rock substrate (foundation). Strong tidal influences create distinct tidal zones. Changing tides cause rocky shore areas to be completely covered with water at certain times of the day. At other times, these same areas can be completely exposed to the air (aerial exposure) and sunlight. These abiotic factors make the rocky shore one of the most physically stressful marine ecosystems. Organisms that live along the rocky shore must be able to tolerate extreme changes in temperature, salinity, moisture, and wave action. Sometimes these changes occur more than two times a day. Tide pools and rocky shores are highly complex and biologically diverse. Numerous marine invertebrates, including urchins, crabs, mussels, barnacles, anemones, sea stars, and sea hares are well-adapted to the extreme conditions of the rocky shore ecosystem. Organisms of the rocky shore face some other threats. They must also protect themselves from terrestrial species like humans, birds, cats, rodents that have access to the rocky shore.

Marine Ecosystem Classification

Open Marine Ecosystems

The first thing many people think of upon hearing the term "marine ecosystem" is the open ocean, which is indeed a major type of marine ecosystem. This category includes types of sea life that float or swim, such as algae, plankton, jellyfish and whales. Many creatures living in the open ocean inhabit the upper layer of the ocean where the sun's rays penetrate. This is known as the euphotic zone and extends to a depth of about 150 meters (500 feet).

Ocean Floor Ecosystems

Marine life not only exists in the open ocean waters, but on its floor as well. Species that live in this ecosystem include certain types of fish, crustaceans, clams, oysters, worms, urchins, seaweed and smaller organisms. In the shallow water, sunlight can penetrate to the bottom. However, at greater depths, sunlight cannot penetrate, and organisms inhabiting this deep water rely on the sinking of organic matter above for survival. Many such organisms are small and generate their own light to find or attract food sources.

Coral Reef Ecosystems

Coral reefs are a special subtype of seafloor ecosystem. Found only in warm tropical waters and at relatively shallow depths, coral reefs are among the most productive ecosystems on the planet. About one-quarter of marine species depend on coral reefs for food, shelter or both. While coral reefs are famous for attracting brightly colored exotic fish, a plethora of other species -- snails, sponges and seahorses, to name a few -- inhabit coral reefs. The reefs themselves are produced by simple animals that build external skeletons around themselves.

Estuary Ecosystems

The term "estuary" typically describes the shallow, sheltered area of a river mouth where freshwater intermingles with saltwater as it enters the sea, although the term can also refer to other areas with flowing brackish waters, such as lagoons or glades. The degree of salinity varies with the tides and the volume of outflow from the river. The organisms inhabiting estuaries are specially adapted to these distinct conditions; hence, the diversity of species tends to be lower than in the open ocean. However, species which generally inhabit neighboring ecosystems may occasionally be found in estuaries. Estuaries also serve an important function as nurseries for many types of fish and shrimp.

Saltwater Wetland Estuary Ecosystems

Found in coastal areas, saltwater wetlands may be considered a special type of estuary, as they also consist of a transition zone between land and sea. These wetlands can be divided into two categories: saltwater swamps and salt marshes. Swamps and marshes differ in that the former are dominated by trees while the latter are dominated by grasses or reeds. Fish, shellfish, amphibians, reptiles and birds may live in or seasonally migrate to wetlands. Additionally, wetlands serve as a protective barrier to inland ecosystems, as they provide a buffer from storm surges.

Mangrove Ecosystems

Some tropical and subtropical coastal areas are home to special types of saltwater swamps known as mangroves. Mangroves may be considered part of shoreline ecosystems or estuary ecosystems. Mangrove swamps are characterized by trees that tolerate a saline environment, whose roots systems extend above the water line to obtain oxygen, presenting a mazelike web. Mangroves host a wide diversity of life, including sponges, shrimp, crabs, jellyfish, fish, birds and even crocodiles.

While the ocean is divided into zones and layers, these are broad categories that do not specify the diversity of ecosystems present. Each layer or zone includes several ecosystems, which have adapted to specific habitats found in those oceanic regions. Marine life can be found from lush shorelines to deep, oceanic trenches.

Oceanic Zones and Layers

The ocean is divided into four major zones: the intertidal, neritic, oceanic and abyssal. The intertidal zone is the area of coastal sea which is affected by tidal changes. This zone contains diverse ecosystems, such as beaches, estuaries and tidal pools. The neritic zone is shallow ocean extending to the edge of the continental shelf, and the oceanic zone is the area located over the abyssal plain. The abyssal zone refers to the vast, dark plains of the ocean basin's floor. It also includes the volcanic rifts of underwater mountain ranges. While the zones are divided like water columns over specific areas of a tectonic plate, the ocean layers are divided based on depth and light regime. The uppermost oceanic layer, called the epipelagic, is followed by the mesopelagic and the bathypelagic in increasing depth; the abyssopelagic is the deepest layer.

Shoreline Ecosystems

Many different ecosystems and communities thrive on the changing shorelines of oceans. Sandy beaches support birds, crustaceans and reptiles, while tidal pools provide temporary refuge for stranded sea creatures and optimum hunting grounds for predators. Estuaries and marshes have a mixture of freshwater and seawater, supporting a diverse community of organisms. These smaller ecosystems are all part of the larger community that inhabits an ocean's shoreline.

Coral Reefs

Coral reefs are formed by dead and living coral. Although these organisms appear plant-like, they are actually tiny animals. Some coral are solitary, but most are colonial and form a larger coral made of individual polyps. The remains of dead coral gradually accumulate to form reefs, which support a wide variety of marine animals, such as:

- Fish
- Octopi
- Eels
- Sharks
- Crustaceans

Mangroves

This ecosystem revolves around mangrove trees, which is a non-taxonomic classification for trees and shrubs that can live in wet, saline habitats. Mangrove ecosystems are found on a quarter of the world's tropical shorelines. This environment is a breeding ground for many species of fish and birds, and is diverse in specialized plant species.

Open Ocean

The open ocean is a broad ecosystem that exists in the light-rich surface layer. The producers for this ecosystem are photosynthetic plankton, which are eaten by fish, rays and whales. Many predators in the open ocean feed on fish and other predators. This ecosystem supports the largest mammal in the world, the blue whale. Ocean currents are an important factor in the life cycles of organisms in the open ocean, bringing nutrient-rich water from other areas.

Deep Ocean

Deep ocean ecosystems are devoid of light and depend on sunken remains and organic materials from the upper oceanic layers. The ocean floor supports various scavengers and their predators, which all benefit from the organic matter drifting down to the floor. The volcanic rifts which form new seabed also support an extremely specialized community of organisms that depend on superheated, smoking vents in the earth's surface. These vents spew out hot water that is rich in minerals. Chemoautotrophic bacteria create energy by oxidizing the sulfur from the vents, and provide food for crab and shrimp species. Tube worms also harbor the energy from chemical reactions to support life, making solar energy absolutely unnecessary for the survival of this ecosystem.

Four Types of Aquatic Ecosystems

Aquatic ecosystems consist of interacting organisms that use each other and the water they reside in or near for nutrients and shelter. Aquatic ecosystems are divided into two major groups: marine, or saltwater, and freshwater, sometimes called inland or non-saline. Each of these can be further subdivided, but the marine types are more typically grouped together than the freshwater ecosystems.

The Largest Ecosystem

Oceans are the largest of the ecosystems, covering more than 70 percent of the Earth's surface. The ocean ecosystem is divided into four distinct zones. The deepest zone of this marine ecosystem, the abyssal zone, has cold, highly-pressurized water with high oxygen but low nutrient levels. Ridges and vents on the ocean floor that emit hydrogen sulfide and minerals are found in this zone. Above the abyssal zone is the benthic zone, a nutrient-rich layer that contains seaweed, bacteria, fungi, sponges, fish and other fauna. Above this is the pelagic zone, essentially the open ocean, which features water with a broad temperature range, surface seaweeds and many species of fish as well as some mammals. The intertidal zone, where the ocean meets land, is covered by water during high tide and is terrestrial during low tide, allowing it to support unique vegetation and animal life.

Rainforests of the Sea

Coral reefs cover only a tiny fraction of the Earth's surface and only a slightly larger percentage of the ocean bottom but support a great deal of diverse aquatic life. Reef-building corals exist only in shallow subtropical and tropical waters. The corals host photosynthesizing algae and get most of their food from these algae, allowing for enough growth to form large structures that create valuable habitat. Rising water temperatures and increasing acidification of water linked to increases in carbon dioxide are the greatest threats coral reefs face. On local levels, over-harvesting of coral and overfishing threatens reefs, as do invasive species and polluted runoff.

Looking at Shorelines

Like coral reefs, estuaries are sometimes grouped with oceans to make up the marine ecosystem. Estuaries occur where saltwater from the ocean and freshwater flowing from rivers or streams meet, creating a unique habitat oriented around water that has a varied salt concentration and has high levels of nutrients resulting from sediments being deposited by rivers or streams.

Lakes and Ponds

Lakes and ponds, water bodies with varied surface areas and volumes, are also known as lentic ecosystems and are characterized by a lack of water movement. Like oceans, lakes and ponds are divided into four distinct zones: littoral, limnetic, pro-fundal and benthic. Light penetrates the uppermost of these, the littoral, which contains floating and rooted plants. The other zones also each play unique roles in the ecosystem.

Flowing Freshwater

Rivers, streams and creeks are classified as lotic ecosystems. These ecosystems are characterized by flowing freshwater, which moves to a larger river, lake or ocean, and is present during part or throughout all of the year. Because of the water's movement, rivers and streams tend to contain more oxygen than their lentic relatives and have host species that are adapted to the moving water.

Wet Soils and Water-loving Plants

Wetlands are freshwater ecosystems characterized by the presence of water, which could be several feet deep or simply saturate the soil, often with seasonal fluctuations. Certain types of soil known as hydric soils that are different than other soils and plant species adapted to wet conditions also characterize wetlands. Wetlands are very important in regulating water levels, filtering water and improving water quality, reducing flood dangers and providing valuable habitat for plants and animals.

Marine Flora and Fauna

Coral reefs and seagrass beds play a vital role in tropical marine environments. In both of these ecosystems, larger algae (macro-algae) play an important part: they provide food for herbivores, and they stabilize the structure of reefs. Algae are also remarkable in that they are responsible for the high productivity that characterizes coral reefs and seagrass beds.

1. Flowering Plants: Seagrasses are the only flowering plants that live under the sea.

2. Green Algae: Green algae, the most diverse group of algae, are common on tropical reefs.

3. Brown Algae: People who live in the temperate zones are familiar with brown algae because of the great size that some species, such as the kelps, attain.

4. Mermaid's Tea Cup Udotea Cyathiformis: It is a green alga having a goblet or funnel-shaped blade, with a sharp junction between the blade and the stalk. It lives on sand at variable depths.

5. Red Algae: There are numerous types of red algae, ranging from filamentous to crustose. Crustose coralline algae are especially important for the structure of coral reefs. These red algae grow as a crust over and between the gaps in coral reefs, and cement the corals together.

6. Marine Plants and Algae on St. Eustatius: St. Eustatius has a rich marine flora, with more than 150 species of macro-algae, and four species of seagrasses. Macro-algae vary tremendously in shape and color, and are found in a range of habitats. Various different species flourish in shallow and deep areas, and on hard substrates as well as in sandy areas. Seagrasses, covered in macro-algae, form dense beds, where turtles feed and other animals seek shelter. However, there is one invasive seagrass species around St. Eustatius which competes with the three native seagrasses. It is important to learn more about how this is impacting the seagrass ecosystems on St. Eustatius.

7. Broadleaf Seagrass Halophila stipulacea.: This invasive seagrass species from the Red Sea has flat leaves up to 5 cm high, growing in pairs. It has slightly toothed leaf edges, with two large scales at the base. It competes with native Caribbean seagrasses.

8. Fan Algae Udotea spp.: The wedge-shaped blade of this species rises from a stem connected to the holdfast. The blade is lightly to moderately calcified. These algae are found in a large range of habitats, from shallow areas down to 50 m.

9. Manatee Grass Syringodium filiforme: Seagrass with thin, cylindrical leaves 2-3 mm in diameter, and up to 30 cm high. Native to the Caribbean. Previously it formed dense seagrass beds, but now is mainly found mixed with the invasive Broadleaf seagrass.

10. Small-leaf Hanging Vine Halimeda goreaui: Forms chains of semi-rectangular blade segments, which are held together by a thin strand. Tends to grow in shaded areas of the reef, often hanging from ledge undercuts, and along walls, down to 80 m.

11. Shaving Brush Algae Penicillus spp.: A green algae with the peculiar shape of a brush. The tip of the brush merges into a short stalk anchored in the substrate. It is abundant in shallow, calm, protected water, but is also seen as deep as 30 m.

12. Green Feather Alga Caulerpa sertularioides: Green alga with several feather-shaped, upright branches up to 20 cm high, arising from horizontal, creeping runners. Often grows in shallow sandy areas, or within seagrass beds.

13. Saw-blade Alga Caulerpa serrulata: Small blades, often twisted or spiralling, with serrated edges. Blades grow upwards from long runners. Green in colour, often with a bluish tint. Grows in shallow, rocky substrates, usually with some sand covering.

14. Leathery Lobeweed Lobophora variegate: This brown algae has three different forms, depending on depth and habitat: fan-shaped, encrusting or ruffled. Light brown to orange in color. It is a very common algae that grows over coral and rocks.

15. Sea Pearl Valonia ventricosa: This alga forms large, thin-walled, round or elliptical sacs. The sacs are single cells that may grow to 10 cm in diameter, among the largest known on Earth.

16. Wireweed species Sargassum spp.: these brown algae can reach several meters in length. They are leathery, tough and grow erect. The axils bear small blades and air bladders.

17. Mermaid's Wineglass Acetabularia schenckii : Blades in the form of small parasols, consisting of fused rays. 3-8 cm high. Heavily calcified and can appear to be green or white. Grows in shallow areas, attached to stones, shells or coral.

18. Broadleaf Gulfweed Sargassum fluitans: A brown alga with short-stalked blades bearing round air bladders with no spine. Does not grow attached to rocks, but instead forms free-floating clumps which drift in the ocean or wash up on the beach.

19. Peacock's Tail Algae Padina spp.: This brown alga forms rounded, thin, undulating blades that curve upward near the edges. The surfaces of the fans are calcified and whitened. Attaches to rocky substrates on shallow reef flats.

20. Red Fan Alga Flahaultia tegetiformans: Inconspicuous alga with fan-shaped blades and a minute stipe, reaches 1-4 cm in diameter. Dark purple-red colour with a golden sheen. Grows in dark caves and in crevices, up to 30 m depth.

21. Y-branched Algae Dictyota spp.: Algae with flat, strap-shaped blades. All species have branches that fork near the end. Generally form mats of leaves that overgrow the substrate. Found in many reef environments, but most common in protected areas.

22. Twig Algae Amphiroa spp: Heavily calcified algae, but with non-calcified joints. Dichotomously branched. Pink or white in colour. Brittle and fragile. Some species grow as tall as 15 cm.

23. Red Lace Alga Martensia pavonia: Lobed blades that form thin, delicate lacework, pale pinkish-blue in colour. Stalks are absent. A common species which often grows attached to other algae.

24. Crustose Coralline Algae: These rock-hard algae grow as calcified crusts on hard substrates. Most species are difficult to identify in the field. Very important in the formation of coral reefs; they act as cement and promote coral growth.

Chapter 3

Physical and Chemical Oceanography

The scientific study of motions, physical properties, conditions and processes within the ocean waters is termed as physical oceanography. Some of the topics studied under physical and chemical oceanography are sea water properties, pressure in ocean water, etc. This chapter has been carefully written to provide an easy understanding of physical and chemical oceanography.

Physical Oceanography

The Physical Oceanography is an essential part of oceanographic analysis. It is the study of physical conditions that are prevailing in the seas and oceans. It deals with all large scale physical processes and their effects that are happening within the oceans. The processes are very dynamic in nature. They involve the water masses which have heterogeneous proportions and dimension. The properties of water masses also vary with space and time. Physical oceanography considers all these aspects in projecting the oceans. This branch has grown due to historical oceanographic explorations and expeditions carried out by several scholars from almost all parts of the world. Very essential aspects of physical oceanography are to be understood first.

The Physical Setting

The oceans are very thin layer of water masses when we consider them on a global scale. The vertical scale is very small when compared to the horizontal scale. Hence, for all graphical representation, the vertical scale is to be exaggerated. The ratio of depth to width of the ocean basins is also very small. This ratio is very important for understanding the ocean currents. The Vertical velocities must be much smaller than the horizontal velocities. Very small amounts of vertical velocities may have a great influence on the water turbulence in these bodies.

Sea-floor Features

Before scientists invented sonar, many people believed the ocean floor was a completely flat surface. Now we know that the seafloor is far from flat. In fact, the tallest mountains and deepest canyons are found on the ocean floor; far taller and deeper than any landforms found on the continents. The same tectonic forces that create geographical features like volcanoes and mountains on land create similar features at the bottom of the oceans.

If you follow the ocean floor out from the beach at the top left, the seafloor gently slopes along the continental shelf. The sea floor then drops off steeply along the continental slope, the true edge of the continent. The smooth, flat regions that make up 40% of the ocean floor are the abyssal plain. Running through all the world's oceans is a continuous mountain range, called the mid-ocean ridge. The mid-ocean ridge is formed where tectonic plates are moving apart from each other, allowing magma to seep out in the space where the plates pulled apart. The mid-ocean ridge system is 80,000 kilometers in total length and mostly underwater except for a few places like Iceland. Other underwater mountains include undersea volcanoes (called seamounts), which may rise more than 1,000 meters above the ocean floor. Those that reach the surface become volcanic islands, such as the Hawaiian Islands. Deep oceanic trenches are created where a tectonic plate dives beneath (subducts) another plate.

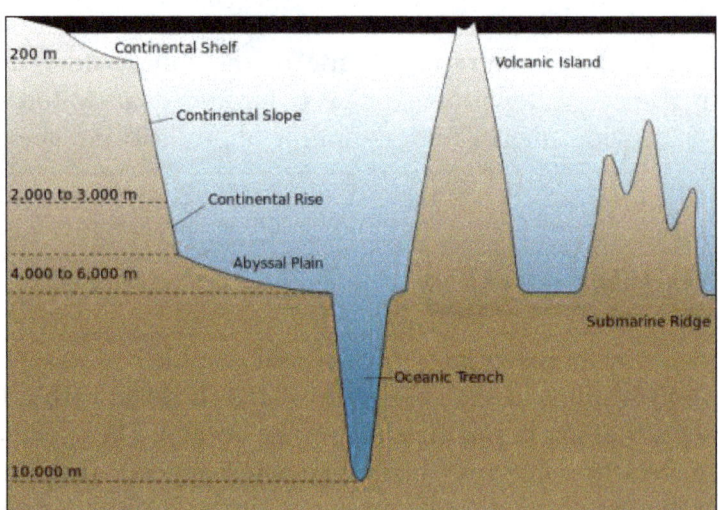

Seawater Properties

One of the most obvious properties of seawater is its salty taste. If you believe that its taste is due to dissolved salt, you are correct, and this is what makes the water of the oceans different from the water of lakes and rivers. Seawater is made up mainly of liquid water (about 96.5 percent by weight) in which chloride (Cl) and sodium (Na) are the dominant dissolved chemicals. The common table salt you use to flavor your food is composed of precisely the same elements.

Basic Chemical Notions

All matter is composed of "building blocks" termed atoms. An atom is the smallest unit of a substance that retains all of its chemical properties. For example, a single atom of hydrogen possesses all of the chemical characteristics of a large collection of hydrogen atoms. If you tried to divide a single hydrogen atom into simpler units, you could do so, but these bits of matter would no longer display the properties of a hydrogen atom. Atoms that are chemically bonded to one another comprise a molecule. Or, if you prefer, a molecule is a chemical substance that can be separated into distinct atoms. Sodium chloride (NaCl) is a molecule of salt that can be separated into a positively charged atom (Na^+) and a negatively charged atom of (Cl^-).

The internal structure of any atom consists of elementary particles that possess mass and electric charge. The center of an atom, the nucleus, is composed of two distinct kinds of particles that contain essentially all of its mass: protons with a positive electric charge and neutrons with no electrical charge. Surrounding the nucleus are collections of orbiting electrons that have little mass and carry a negative electrical charge. The electron orbits are not randomly distributed around the nucleus but are confined to discrete levels termed electron shells. All elements (except for one type of hydrogen that does not have neutrons) contain protons, neutrons, and electrons, and differ from one another because of the number and structural arrangement of these fundamental subatomic particles. A stable atom of an element is electrically neutral, indicating that the positive charges from the protons are balanced by the negative charges from the electrons.

Hydrogen possesses one proton in its nucleus; the positive charge from its single proton is neutralized by the negative charge from its one orbiting electron. Oxygen, on the other hand, contains eight protons in its nucleus, balanced electrically by eight orbiting electrons. So hydrogen atoms consist of a single proton and a single electron, and oxygen atoms of eight protons and eight electrons. If electrons are either added or removed from any single atom, the atom is no longer electrically balanced. Atoms with more electrons than protons have a net negative charge. Atoms with more protons than electrons have a net positive charge. An atom with either a positive or a negative charge is called an ion.

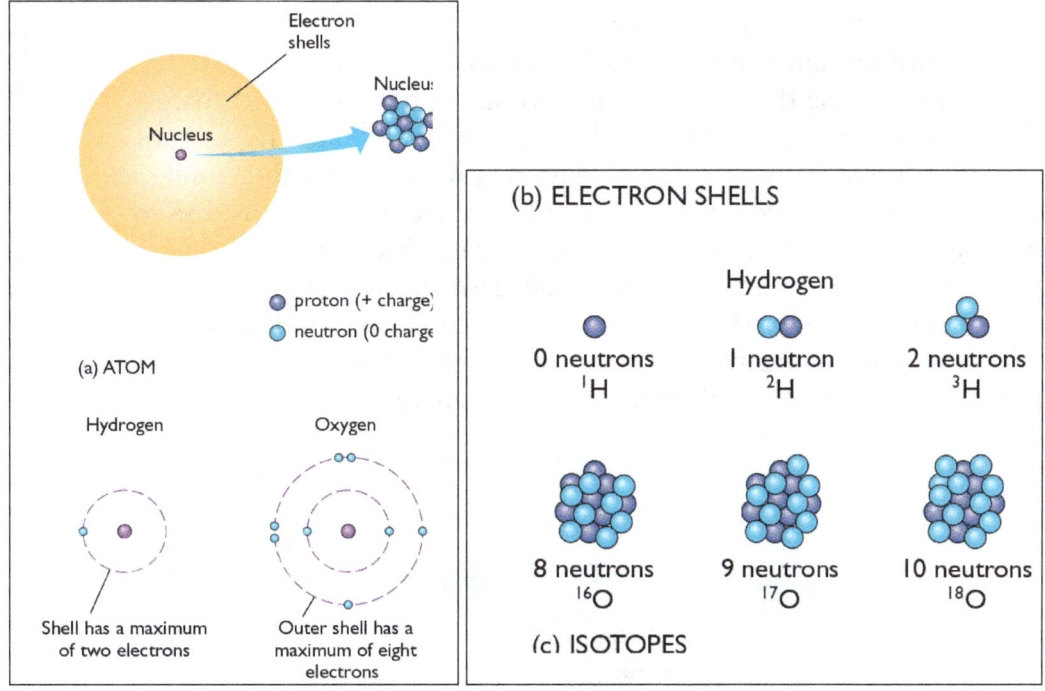

Figure: Atomic structure.

Figure shows (a) A simplified version of an atom depicts a nucleus of protons and neutrons, surrounded by shells of orbiting electrons. (b) A depiction of electrons within the shells of hydrogen (one electron) and oxygen (eight electrons) atoms. (c) The isotopes of an element have a variable number of neutrons. Hydrogen isotopes have zero, one, or two neutrons; oxygen isotopes have eight, nine, or ten neutrons. For example, when NaCl dissolves in water, it separates into the ions Na$^+$ and Cl$^-$. The charge of an ion is the single most important reason for its ability to bond with other elements.

The story does not end there though. Although the number of protons is fixed for any element, the quantity of neutrons in its nucleus can vary. Because the neutron carries no electrical charge but has mass, variations in the number of neutrons change the weight of the element, but not its basic chemistry. Atoms of the same element that differ in weight due to variable numbers of neutrons are called isotopes. Hydrogen has three isotopes. Each isotope contains a single proton, but zero, one, or two neutrons. Oxygen similarly has three isotopes containing eight, nine, or ten neutrons in its nucleus. The most abundant hydrogen isotope has zero neutrons and the most abundant oxygen isotope has eight.

Basic Physical Notions

Another important notion that we must consider in order to understand the behavior of water is the physical concept of heat, the property that one measures with a thermometer and that results from the physical vibrations of atoms and molecules. These physical vibrations represent energy of motion, called kinetic energy. The more heat in a material, the greater the agitation of its atoms and molecules.

A block of ice consists of an orderly, rigid arrangement of water molecules that are held firmly in place by strong electrical bonds between the molecules. Although an ice cube appears inert to the naked eye, it is not inert on an atomic scale. Its molecules are vibrating back and forth even though they are locked together into a crystalline framework. If we add heat to the ice cube, the molecules vibrate faster and move farther back and forth in the crystal. Above 0°C, the melting temperature of ice, the back-and-forth motions of the molecules are so vigorous that they exceed the strength of the electrical bonds holding the molecules in place in the crystal. At that temperature, the crystal's structure disintegrates (melts), and the solid ice becomes liquid water. Liquids are loose aggregates of molecules that are in contact, but are free to move relative to one another, unlike the fixed molecules in a solid. As more heat is added to the liquid water, its temperature rises as its molecules vibrate more energetically. Some of these molecules contain so much kinetic energy that they escape from the liquid surface and become a gas; this process is called evaporation. At 100°C, the boiling temperature of water, all of the molecules are highly energized and thus are vaporized (converted into gas). Free gas molecules move independently of one another between collisions, and are the least ordered of the three states of matter because of their very high kinetic energy.

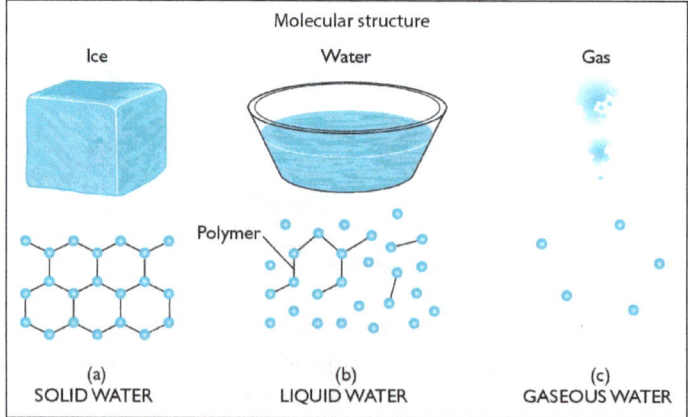

Figure: States of matter.

The kinetic theory of heat has important implications for measuring temperature and understanding the density of materials, whether they be solids, liquids, or gases. Imagine a thermometer, which is nothing more than a narrow glass tube filled with mercury. If we wish to determine the temperature of a water sample, we stick the thermometer in the liquid. The mercury in the tube rises and eventually stops.

Water occurs in three states, which depend on temperature and pressure. (a) Solid water (ice) consists of ordered molecules that are tightly bonded to one another. (b) Liquid water consists of molecules that move relative to one another. Polymers are bits of crystalline structure that can exist in liquid water near its melting temperature. (c) Gaseous water (gas) is made up of independently moving molecules.

It is then an easy matter to read the temperature of the water off the thermometer's scale. But what exactly does the rise of mercury in the tube represent in a physical sense? Well, it's quite simple, provided you grasp the kinetic theory of heat. Recall that the water molecules are vibrating at a rate that depends on the water's temperature. When the thermometer is placed in the liquid, water molecules strike the tube; these collisions add energy to the molecules of the tube, so they vibrate faster. This in turn transfers kinetic energy to the atoms of mercury. The atoms of mercury begin to vibrate vigorously and collide with one another harder and harder, forcing them on the average farther apart than they were before the addition of kinetic energy from the water. Therefore, the mercury expands and rises in the tube of the thermometer, which is calibrated precisely to read temperature.

Temperature controls another fundamental property of matter called density—the amount of mass contained in a unit volume, expressed as grams (mass) per cubic centimeter (volume), that is, g/cm^3. The more mass that is contained in a cubic centimeter, the denser is the material. The amount of heat contained in a substance determines how vigorously the molecules in that substance vibrate and collide with one another. The harder they vibrate, the farther apart they tend to be, which controls directly the amount of mass that is contained in a unit volume. This means that 10°C water is denser than 15°C water and that warm air is less dense than cold air at the same pressure.

Now that we have a solid conceptual understanding of the structure of atoms, of the kinetic theory of heat, and of temperature and density, we can examine the chemical and physical properties of water in general, and seawater in particular.

The Water Molecule

Most people know that the chemical formula for water is H_2O. This formula means that water consists of two atoms of hydrogen (H) that are chemically bonded to one atom of oxygen (O). However, despite its simple chemical composition, water is a complex substance with truly remarkable physical properties. For example, the melting and boiling temperatures of water are much higher than expected when compared with chemically related hydrogen compounds. This is fortunate; otherwise, water would be able to exist only as a gas at the temperatures that prevail at the Earth's surface. Consequently, the oceans could not have formed, and life could not have developed. In fact, H_2O is the only substance that can coexist naturally as a gas, a liquid, and a solid on the Earth's surface. Therefore, it is not surprising to discover how fundamental it is to all forms of life.

Water also has an unusually high heat capacity and tremendous solvent power. Heat capacity is defined as the quantity of heat required to raise the temperature of 1 gram of a substance by 1°C. More energy is required to raise the temperature of a substance with high heat capacity than one with low heat capacity. In other words, adding the same amount of heat will raise the temperature of a substance with a low heat capacity to a greater degree than one with a high heat capacity. The high heat capacity of water explains why so much energy is required to heat water.

In addition to its unusually high heat capacity, the capability of liquid water to dissolve material, its solvent power, is unsurpassed by any other substance. In fact, chemists refer to water as the "universal solvent," meaning that virtually anything can be dissolved to some extent in liquid water.

Water possesses other unusual properties as well, but the case for this substance's chemical uniqueness should be clear. To account for water's singular properties, we must examine the physical structure of the water molecule, H_2O. The molecule's asymmetrical shape is as important as the chemical identities of the two elements H and O. The two hydrogen atoms, rather than being attached symmetrically to either side of the oxygen atom, are separated from each other by an angle of 105 degrees for water in the liquid and gaseous states. This molecular architecture resembles a mouse's "head" with hydrogen "ears." The hydrogen and oxygen atoms share electrons, and this covalent bond creates the water molecule H_2O. Each hydrogen atom within the molecule possesses a single positive charge; the oxygen atom, a double negative charge. As such, the H_2O molecule is electrically neutral because of the balance between the positive and negative charges.

However, the structural asymmetry of the molecule imposes a slight electrical imbalance because the positive hydrogen atoms are bonded to one end of the oxygen atom and the electrons associated with that atom are concentrated on the other side. This gives rise to a dipole (two-pole) structure. Thus, there is a residual positive charge at the hydrogen end of the molecule and a residual negative charge at the oxygen end, despite the overall electrical neutrality of the water molecule. Consequently, liquid water is not merely a collection of freely moving molecules. Rather, its dipole structure causes the negative end of the molecule to be attracted to and become electrically bonded to the positive end of a nearby water molecule. This electrostatic bonding, called hydrogen bonding, produces irregular chains and clusters of H_2O molecules. The size of water-molecule clusters decreases with increasing temperature. Although hydrogen bonding is only 4 percent as strong as the covalent bonds that hold the atoms of the molecule together, it is directly responsible for many of water's extraordinary physical properties.

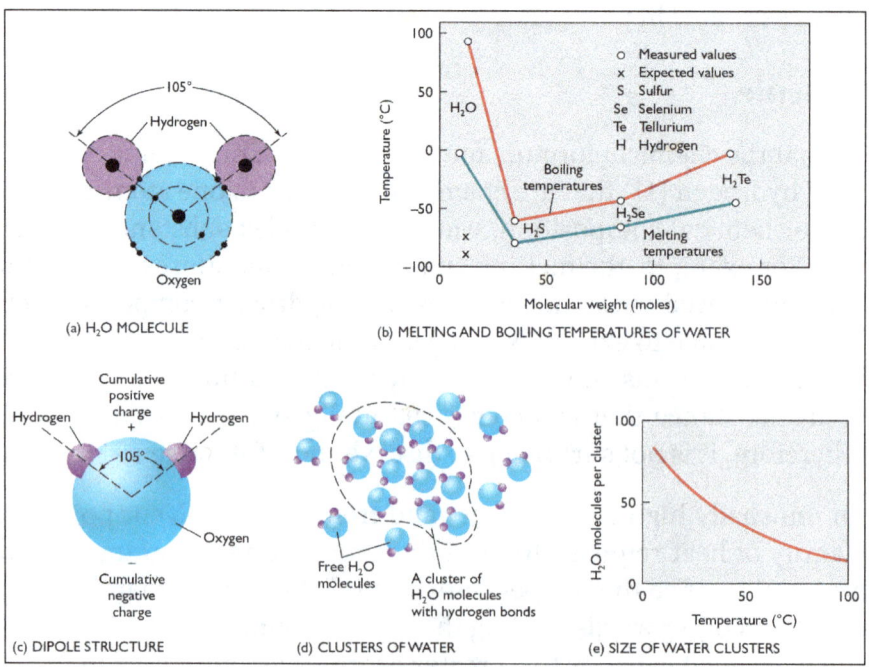

Figure: The water molecule.

(a) The chemical bonding of two hydrogen atoms and one oxygen atom produces a water (H_2O) molecule. (b) The observed (o) melting and boiling temperatures of water are much higher than theory (x) indicates. (c) The two small hydrogen atoms are separated from each other at their points of attachment to the large oxygen atom by an angle of 105 degrees, so that the molecule resembles the familiar caricature of a mouse's head. This structure creates a dipolar molecule with a residual positive charge at one end and a residual negative charge at the other end. (d) A cluster of H_2O molecules with hydrogen bonds is contrasted here with free H_2O molecules. (e) The size of the clusters decreases with increasing temperature.

The higher-than-expected melting and boiling temperatures of water depend directly on the dipole structure of the H_2O molecule. More energy is required than expected to vaporize liquid water and to melt ice because hydrogen bonds that link H_2O molecules to H_2 molecules must first be broken before the solid can melt and the liquid can vaporize. This is also the reason for water's high heat capacity. When heat is added to water, only a fraction of this energy is actually used to increase the vibrations of the molecules, which would be detected as a rise of temperature. Much of the added heat is used to break hydrogen bonds that link the H_2O molecules into irregular clusters. Thus, as water is heated, its temperature rises slowly relative to the amount of energy used. Conversely, when cooled, water releases more heat than expected from the decrease in its temperature.

The unusually high heat capacity of water prevents extreme variations in the temperature of the oceans and explains why the climates of coasts and islands experience less extreme temperature variations than those of land located far from the ocean or large lakes. During the summer, large bodies of water absorb solar heat, helping keep air temperatures cool. During the winter, large lakes and the ocean radiate great quantities of stored heat to the atmosphere as their water cools, warming the adjacent shoreline.

The high solvent power of liquid water likewise depends on the dipole structure of its molecule. Sodium and chloride are the most common elements dissolved in seawater. Sodium is a positively charged atom (Na^+) called a cation. In contrast, chloride is a negatively charged atom (Cl^-) called an anion. When these two ions come into contact, they are attracted to each other because of their opposing charges and can be held together by ionic bonds to form halite (rock salt or common table salt). When halite crystals are put in water, the negatively charged end of the H_2O dipole dislodges sodium ions (Na^+) from the solid, and the positive end of the H_2O molecule tears off chloride ions (Cl^-). Dissolution (the noun form of the verb to dissolve) continues until either the entire crystal of halite is gone or the volume of water can no longer accommodate more ions of salt because it is saturated, meaning there is physically no more "room" for the sodium and chloride in the water. (By way of an analogy, think about what happens when you keep adding spoonful after spoonful of sugar to your cup of coffee.) In solution, the sodium and chloride ions are surrounded by water molecules. This keeps the cations separated from the anions, a process known as hydration. In other words, water acts as a solvent by preventing the chemical recombination of $Na+$ and Cl^- to form the solid halite.

The density of water is yet another unusual property of this familiar substance. Ice floats on water! All other solids sink in their own liquids. It's hard to imagine a solid bar of steel floating in a vat of molten steel. Once again, water behaves in a peculiar way.

At the freezing point, solids crystallize from liquids because the thermal vibrations of molecules are low enough so that chemical bonding can occur and a crystal forms. The loose assortment of molecules in the liquid is reconstructed into a rigid solid. Because the molecules in the solid vibrate less than do the molecules in the liquid, they are more tightly packed and denser in the solid than in the liquid state. Therefore, solids, because of their higher density, sink in their own liquids. Water does not behave in the same way because ice molecules are arranged into an open crystal framework, whereas liquid water molecules are packed into snug clusters by the hydrogen bonds between molecules. Also, the angle of separation between the two hydrogen atoms, which is 105 degrees in liquid and gaseous water, expands to 109.5 degrees in ice. This makes ice about 8 percent less dense than water. The H_2O molecules in ice are ordered into a porous hexagon (a six-sided structure) by the hydrogen bonds between oppositely charged ends of neighboring molecules.

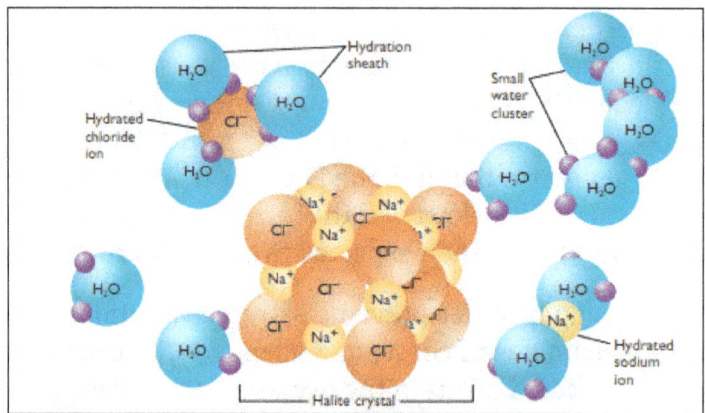

Figure: Halite (rock salt)-The dipole structure of the H_2O accounts for its unsurpassed properties as a solvent. Na^+ and Cl^- ions are dislodged from the halite crystal by, respectively, the negatively and positively charged ends of the water dipole.

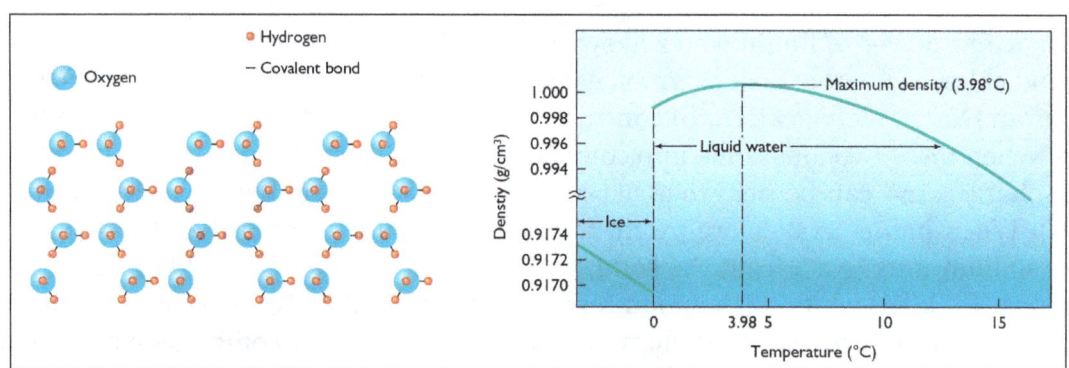

Figure: Properties of water-(a) The open network of the hexagonal structure of ice crystals is shown here. (b) Because of the open-crystal structure of solid water, ice is less dense than liquid water. Water attains its maximum density at a temperature of 3.98°C; polymers exist in liquid water colder than this temperature.

When ice is warmed to 0°C, it begins to melt as thermal vibrations of the molecules cause the crystalline structure to break apart. Because of hydrogen bonds, the freed H_2O molecules in the liquid become more closely packed together than they were in the solid, resulting in an increase of density (more mass per unit volume). Water does not, however, reach its maximum density until it is warmed to 3.98°C because loose aggregates of molecules that resemble the open crystalline structure of ice, termed polymers, persist in water cooler than that critical temperature. Above 3.98°C, the density of water decreases with increasing temperature, as expected. Below 3.98°C, however, the density of water decreases as temperature decreases until it freezes into ice.

What we have learned so far about seawater is that it consists dominantly of H_2O, a chemical substance that can occur in a solid, liquid, and gaseous state at the temperatures and pressures that exist at the Earth's surface. Although we find water everywhere on the planet, its unusual chemical and physical properties should not be taken for granted; it is these very properties that have enabled life to appear and evolve through geologic history. We can now proceed to examine seawater's other chemical ingredients.

The Solutes of Seawater

Chemical analyses of samples from all over the world show that seawater consists of a small quantity of salt dissolved in water. The salt occurs as charged particles, cations and anions that are dispersed among the molecules of liquid water. Seawater is a chemical solution. The dissolving agent, the liquid water, is the solvent and the dissolved substances, the salt ions, are the solute. Seawater also contains minor to trace amounts of dissolved metals, nutrients, gases, and organic compounds of seemingly infinite variety.

Before examining the chemistry of seawater solutions, we need to determine the amount of material that is dissolved in seawater so that we can compare samples taken from different parts of an ocean or, for that matter, from different oceans. We can taste seawater samples and say qualitatively that this sample tastes "saltier" than this other one. But this is a rather subjective technique, and a scientist needs to know exactly and precisely how salty a parcel of water is. Oceanographers specify the concentration of a solute in seawater in units called parts per thousand, represented by either the abbreviation ppt or, as preferred by marine scientists, the symbol ‰. Although oceanographers also express salinity as a dimensionless unit in terms of PSS78—Practical Salinity Scale 1978—we will use ppt. Average or "normal" seawater has a salinity of about 35‰. This means that the dissolved salt occurs in a concentration of 35 parts per thousand (ppt). That is, the salt comprises 3.5 percent (divide 35 by 1,000, and convert it to a percentage by multiplying it by 100) of the sample, the rest (96.5 percent or 965 parts per thousand) being H_2O molecules. (Notice that 965 + 35 = 1,000.) This signifies that a volume of seawater weighing 1,000 grams (or 1 kilogram) with a salinity of 35‰ contains 35 grams of solute. Obviously, a 100-gram sample of seawater with a salinity of 35‰ contains 3.5 grams of dissolved salts.

As you know, because of its high solvent power (ability to dissolve), ocean water dissolves many types of chemicals, and most are found in minute quantities. All these solutes can be grouped into five broad categories: major constituents, nutrients, gases, trace elements, and organic compounds.

Major Constituents

In terms of quantity, the primary solutes in seawater are cations and anions. By weight, chloride (Cl^-) and sodium (Na^+) together comprise more than 85.65 percent of all the dissolved substances in seawater. When these two ions bond chemically into a solid, they form halite and give seawater its most distinctive property—its saltiness. Surprisingly, the six most abundant ions—chloride (Cl^-), sodium (Na^+), sulfate (SO_4^{2-}), magnesium (Mg^{2+}), calcium (Ca^{2+}), and potassium (K^+)—make up over 99 percent of all of seawater's solutes. The addition of five more solutes to the list—bicarbonate (HCO_3^-), bromide (Br^-), boric acid (H_3BO_3), strontium (Sr^{2+}), and fluoride (F^-)—elevates the quantity of dissolved ingredients in seawater to 99.99 percent. This means, of course, that everything else dissolved in seawater occurs in trace amounts and collectively comprises only 0.01 percent! But what appears to be insignificant cannot be ignored because, even in tiny quantities, many of these chemicals are absolutely critical for life in the ocean. Because the concentrations of these major constituents in seawater vary little over time at most localities, they are described as conservative ions of the ocean.

Nutrients

Nutrients are essential for plant growth, as anybody who has fertilized a lawn or garden knows. All plants, including those that live in the ocean, convert nutrients into food (organic compounds such as sugar) by photosynthesis. Nutrients in seawater are compounds that consist primarily of nitrogen (N), phosphorous (P), and silicon (Si). Representative concentrations of these nutrients in the ocean are listed in table below; the concentrations are specified in parts per million (ppm). Most plants cannot use elemental nitrogen and phosphorus and so satisfy their nutrient needs by absorbing phosphate (PO_4^{3-}) and nitrate (NO_3^-). Silicon is used by important groups of microscopic plants (diatoms) and animals (radiolaria) to precipitate silica (SiO_2) shells around their fragile cells. Because of biological uptake and release, the concentrations of nutrients in seawater, as on land, vary from place to place and over time at any one place. Thus, oceanographers refer to these substances as non-conservative ions of seawater, signifying that levels of these substances are not constant in water but vary over time and from place to place.

Table: Major solutes in sea water.

Salt Ion	Ions in Seawater (‰)	Ions by Weight (%)	Cumulative (%)
Chloride (Cl^-)	18.980	55.04	55.04
Sodium (Na^+)	10.556	30.61	85.65
Sulfate (SO_4^{2-})	2.649	7.68	93.33
Magnesium (Mg^{2+})	1.272	3.69	97.02
Calcium (Ca^{2+})	0.400	1.16	98.18
Potassium (K^+)	0.380	1.10	99.28
Bicarbonate (HCO_3^-)	0.140	0.41	99.69
Bromide (Br^-)	0.065	0.19	99.88
Boric acid (H_3BO_3)	0.026	0.07	99.95
Strontium (Sr^{2+})	0.013	0.04	99.99
Fluoride (F^-)	0.001	0.00	99.99
Total	34.482	99.99	99.99

Near-surface Nutrient Concentrations in Seawater	
Nutrient Element	Concentration (ppm)
Phosphorus (P)	0.07
Nitrogen (N)	0.5
Silicon (Si)	3

Gases

Listed in order of decreasing abundance, gases in seawater include nitrogen (N_2), oxygen (O_2), carbon dioxide (CO_2), hydrogen (H_2), and the noble gases argon (Ar), neon (Ne), and helium (He). Nitrogen and the three noble gases are inert (unreactive) and rarely involved directly in plant photosynthesis. In contrast, levels of dissolved O_2 and CO_2 are greatly influenced by photosynthesis and respiration of organisms. Therefore, they vary greatly in space and time depending on the activities of plants and animals and are regarded as non-conservative.

Trace Elements

Trace elements are all chemical ingredients that occur in minute (trace) quantities in the oceans. Most trace elements, such as manganese (Mn), lead (Pb), mercury (Hg), gold (Au), iodine (I), and iron (Fe), occur in concentrations of less than 1 ppm (part per million). Many occur in quantities of less than 1 part per billion (ppb) and even at 1 part per trillion. These low concentrations make certain trace elements difficult and sometimes even impossible to detect in seawater. However, despite their extremely low concentrations, trace elements can be critically important for marine organisms, either by helping to promote life or by retarding or killing life (toxicity).

Table: Examples of trace elements in seawater.

Trace Element	Concentration (ppb)*
Lithium (Li)	170
Iodine (I)	60
Molybdenum (Mo)	10
Zinc (Zn)	10
Iron (Fe)	10
Aluminum (Al)	10
Copper (Cu)	3
Manganese (Mn)	2
Cobalt (Co)	0.1
Lead (Pb)	0.03
Mercury (Hg)	0.03
Gold (Au)	0.004

Organic Compounds

Organic compounds are large, complex molecules produced by organisms. They include substances such as lipids (fats), proteins, carbohydrates, hormones, and vitamins. Typically, they occur in low concentrations and are produced by metabolic (physical and chemical processes in the cell of an organism that produce living matter) and decay processes of organisms. For example, vitamin complexes are vital for promoting the growth of bacteria, plants, and animals, as shown by the control that thiamine and vitamin B12 have on the growth rate, size, and number of microscopic plants grown in laboratory experiments.

Now that we have a general understanding of the chemical makeup of seawater—a solution of mainly water with some salts, and tiny quantities of nutrients, gases, trace elements, and organic compounds—we can proceed to examine the nature of salinity and its effect on the properties of water, as well as the factors that control the saltiness of the ocean.

Table: Quantities of gas in air and seawater.

Gas	In Dry Air (%)	In Surface Ocean Water (%)	Water–Air Ratio
Nitrogen (N_2)	78.03	47.5	0.6
Oxygen (O_2)	20.99	36.0	1.7
Carbon dioxide (CO_2)	0.03	15.1	503.3
Argon (Ar), hydrogen (H_2), neon (Ne), and helium (He)	0.95	1.4	1.5

A simple way to determine salinity is to evaporate water from a container of seawater and then compare the weight of the solid residue left behind in the bottom of the container—the salts—to the weight of the original sample of seawater. Unfortunately, the method is neither precise nor accurate because salt crystals hold on to variable amounts of H_2O molecules, and that affects the weight of the salt residue. In order to compare accurately salinity data gathered from many parts of the ocean and measured in many different laboratories and ships, chemists have adopted a standardized and what seems to non-chemists to be a rather cumbersome definition of salinity: the total mass expressed in grams of all the substances dissolved in 1 kilogram of seawater, when all the carbonate has been converted to oxide, all the bromine and iodine have been replaced by chlorine, and all organic compounds have been oxidized at a temperature of 480°C. Because we are not chemical oceanographers, we can simplify the definition of salinity as follows to suit our more general purpose: the total weight in grams of dissolved salts in 1 kilogram of seawater expressed as ‰ (parts per thousand).

Principle of Constant Proportion

Salinity determinations from the world's oceans have revealed an important, unexpected finding. Although salinity varies quite a bit because of differences in the total amount of dissolved salts, the relative proportions of the major constituents are constant. In other words, the ratio of any two major constituents dissolved in seawater, such as Na^+/K^+ or Cl^-/SO_4^{2-}, is a fixed value, whether the salinity is 25, 30, 35‰, or whatever. To put it in more familiar terms, let's imagine that the ratio of females to males in a population is ¼ (1 female for every 4 males) and that this ratio never changes regardless of population size. This means that the total number of people in the population can vary, but the relative proportion of females to males does not change. In other words, the ratio of females to males is constant and is independent of population size. Just so, the ratio of any two major salt constituents in ocean water is constant and is independent of salinity.

This important discovery, made during the Challenger expedition, is termed the principle of constant proportion or constant composition, and was a major breakthrough in determining salinity of seawater in a rapid, accurate, and economical manner. In theory, all that need be done to quantify salinity is to measure the amount of only a single major ion dissolved in a sample of seawater because all the other major constituents listed in table occur in fixed amounts relative to that ion. Chemists chose to measure Cl^- for determining the salinity of seawater because it is the most abundant solute in seawater and its concentration is easily determined.

Today, oceanographers rely on a variety of methods, including the electrical conductivity of seawater, to make routine determinations of salinity. The electrical conductivity of a solution is its ability to transmit an electrical current directly proportional to the total ion content of the water at a given temperature, which, of course, is its salinity. A salinometer indirectly measures salinity by measuring the electrical conductivity of seawater. Its calibrations allow oceanographers to determine salinity directly and quickly by inserting an electrical probe into the water, a very simple matter compared with the rather laborious chemical determination of chlorinity.

Factors that Regulate the Salinity of Seawater

There is a lot of dissolved salt in the ocean. Rivers disgorge huge volumes of fresh water into the ocean every year. Chemical analyses of water samples from rivers all over the world indicate that they contain a variety of dissolved substances in concentrations of ppm. Rivers have a dissolved load of chemicals because of the chemical weathering of rocks on the land. These rocks are made up of an assemblage of minerals composed predominantly of the elements silicon, aluminum, and oxygen. Acidic water breaks down these rocks into their component elements. When carbon dioxide (CO_2) is dissolved in water, it reacts with H_2O molecules to produce H_2CO_3, a weak acid called carbonic acid. In turn, this acid separates into hydrogen (H^+) and bicarbonate (HCO_3^-) ions. The specific chemical reactions are reversible, are quite simple, and are represented by:

$$H_2O + CO_2 \rightleftharpoons H_2CO_3 \rightleftharpoons H^+ + HCO_3^-$$

Table: Sources and sinks of some sea water components.

Chemical Component	Sources	Sinks
Chloride (Cl^-)	Volcanoes River influx	Evaporative deposition as NaCl (rock salt). Net air transfer. Pore-water burial.
Sodium (Na^+)	River influx	Evaporative deposition as NaCl (rock salt). Net air transfer. Cation exchange with clays. Basalt-seawater reactions. Pore-water burial.
Potassium (K^+)	River influx Volcanic-seawater reactions (high temperature)	Uptake by clays. Volcanic-seawater reactions (low temperature).
Calcium (Ca^{2+})	River influx Volcanic-seawater reactions Cation exchange	Biogenic secretion of shells. Evaporitic deposition of gypsum ($CaSO_4 \cdot 2H_2O$). Precipitation as calcite.

Silica (H_4SiO_4)	River influx	Biogenic secretion of shells.
	Basalt-seawater reactions	
Phosphorus (HPO_4^{2-}, PO_4^{3-}, H_2PO^{4+}, organic P)	River influx	Burial as organic P.
	Rainfall and dry fallout	Adsorption on volcanic ferric oxides.
		Formation of phosphorite rock.

Notice that the reaction yields free ions of H^+, which because of their small size and high chemical reactivity; replace cations such as Na^+ and K^+ that are bound to minerals in rocks. The amount of H^+ is a measure of the acidity of the water. This process—the bathing of rocks in acidic water—slowly weathers minerals, releasing ions, which go into solution and become part of a river's dissolved chemical load.

Table: Dissolved substances in river water.

Substance	Concentration (ppm)	Concentration (%)
Bicarbonate/carbonate (HCO_3^-/CO_3^{2-})	58.8	48.7
Calcium (Ca^{2+})	15.0	12.4
Silica (SiO_2)	13.1	10.8
Sulfate (SO_4^{2-})	1.2	9.3
Chloride (Cl^-)	7.8	6.5
Sodium (Na^+)	6.3	5.2
Magnesium (Mg^{2+})	4.1	3.4
Potassium (K^+)	2.3	1.9
Nitrate (NO_3^-)	1.0	0.8
Iron aluminum oxide [$(Fe, Al)_2 O_3$]	0.9	0.8
Remainder	0.3	0.3

Let's firm up weathering by examining the chemical breakdown of an actual mineral, orthoclase ($KAlSi_3O_8$), the common potassium-bearing feldspar of granite. The chemical reaction is,

$$\underset{\text{(orthoclase)}}{2KAlSi_3O_8} + 2H^+ + H_2O \rightarrow 2K^+ + \underset{\text{(kaolinite)}}{Al_2Si_2O_5(OH)_4} + \underset{\text{(dissolved silica)}}{4SiO_2}.$$

This formula indicates that the mineral orthoclase in the presence of acidic water (H^+) is broken down into potassium ions (K^+), silica (SiO_2), and aluminum silicates. The latter compound bonds with H_2O, forming another mineral, the clay kaolinite. Rivers transport these materials to the ocean in two distinct states: as a dissolved load (K^+ and SiO_2) and as a suspended load (particles of kaolinite). The dissolved K^+ and SiO_2 contribute to the salinity of seawater; the clay accumulates as sediment on the ocean floor. So what was a solid mineral in rocks on land become, by the processes of chemical weathering and transport, dissolved salts in the ocean and mud on the sea bottom.

How much mass is actually added to the ocean by the river influx of dissolved matter? The first response by most people is that the amount could not be much because the concentration of solutes in rivers is low (that's why freshwater doesn't taste salty), and it's dissolved, so you can't see it (how can anything that can't be seen amount too much?). It turns out that the annual river input of material in solution to the oceans is somewhere between 2.5×10^{15} and 4×10^{15} grams. True, 1015 (10 multiplied by itself fifteen times) seems to be quite a big number. But how big is it, really? This rate of influx, 10^{15} grams per year, is about equivalent to the mass of mud supplied each year to the oceans by the rivers of North America, South America, Africa, and Europe combined! Although dissolved material is invisible to the naked eye, it represents a major annual input of mass to the oceans, all of it derived by the chemical weathering of rocks on land. In addition to the supply from rivers, the Earth is degassing. This means that volcanoes on the crests of spreading ocean ridges and in the volcanic arcs of subduction zones spew large quantities of cations (including Ca^{2+} and K^+) and anions (including SO_4^{2-} and Cl^-) into the water column, although the exact amount of this input has yet to be determined reliably.

The fossil record and sedimentary rocks themselves indicate that oceans have existed on the Earth for at least as long as 3.4 billion years. Geochemical data indicate that the salinity of the oceans has changed little over the past 1.5 billion years. This constant ocean salinity despite the tremendous annual supply of dissolved chemicals to the oceans by rivers can only mean that on average a similar quantity of salt must be removed from the oceans each year. Otherwise, the salinity of the world's oceans would have increased over geologic time. This balance between inputs and outputs of salt to the ocean is called steady-state equilibrium.

Oceanographers refer to inputs of material as sources and their outputs as sinks. We've already identified several of the principal sources of the salt ions to the ocean. The removal of salt occurs by both inorganic and organic processes. Evaporation is an excellent example of an inorganic process and, as discussed in the boxed feature, "Desalination," is a technique for producing drinkable water from seawater. In arid climates, evaporation rates are high. Evaporation removes water from the ocean but not the dissolved salt ions. This indicates that, with time, the concentration of salt will rise by the evaporation of water molecules, creating brine, or a very salty solution. The Dead Sea and the Great Salt Lake, both in arid settings, are fine examples of this very process. As more and more water molecules are evaporated from the water, the solution eventually becomes saturated, which means that the solution is holding as much material in a dissolved state as it can for the temperature and pressure conditions of the water. The removal of more water creates a supersaturated solution (a solution containing a quantity of dissolved ions that exceeds the theoretical saturation value). This leads to the precipitation of evaporite minerals from seawater, notably halite (NaCl) and gypsum ($CaSO_4 \cdot H_2O$). Precipitation of evaporite minerals from seawater represents a sink because dissolved salt ions are being removed from the ocean to form sedimentary deposits on the seafloor.

Wind also blows onshore a large amount of sea spray, which on evaporation forms a coating of salt on land—as anybody who wears glasses and lives by the seashore can attest. In addition, freshly extruded basalt lavas on the ocean floor are quite reactive and extract dissolved ions, such as Mg^{2+} and SO_4^{2-}, from the seawater that comes in contact with the hot lava. Finally, adsorption (the "sticking" of ions to a surface) of cations like K^+ and Mg^{2+} by certain clay minerals in the ocean and the formation of hydrogenous minerals, such as ferromanganese nodules, remove a large, unknown quantity of ions from the sea. All of these represent chemical sinks for ions dissolved in seawater.

Organisms help maintain the steady-state equilibrium of the ocean's salinity as well. We already know that diatoms have silica shells and foreams carbonate shells that are precipitated from the uptake of Si^{4+} and Ca^{2+}, respectively, from seawater. Once these organisms die, their hard parts may settle to the sea bottom to form deep-sea oozes. Also, many species of animals extract certain chemical substances that are dissolved in seawater. Some of these chemicals are concentrated in fecal pellets that sink to the ocean floor and become incorporated in sediment.

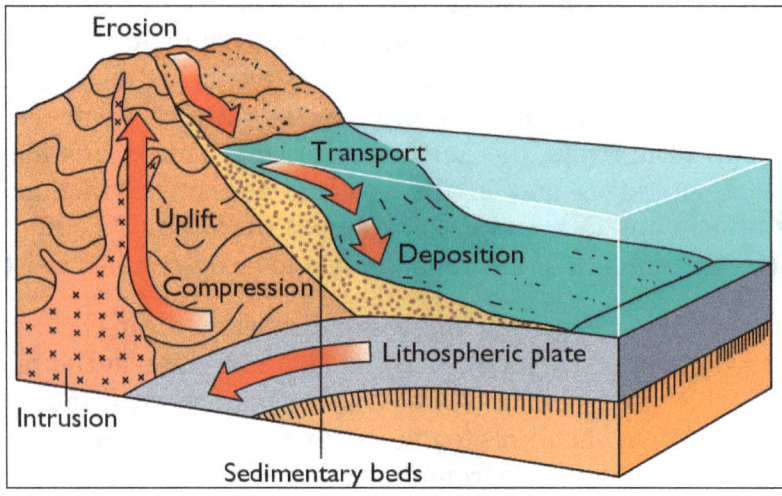

Figure: Sedimentary cycle.

Over geologic time, mountains are leveled by rivers. The weathering products are dispersed into the ocean and collect on the sea bottom, forming sedimentary beds. Eventually, these accumulations of sediment are deformed and uplifted into mountain ranges by plate tectonics. Then a new cycle of river erosion begins.

The salt ions dissolved in ocean water are derived largely from the weathering and erosion of rocks on land. Because the average chemical composition of seawater has remained remarkably stable (in a steady state) over geologic time, the inputs of salt must be balanced by the outputs. The river-supplied ions remain in ocean water for a long time, but eventually are extracted by inorganic and organic processes and become part of the ocean's sedimentary record. These sediments, as they are buried, become cemented into sedimentary rocks and eventually are subducted along the colliding boundaries of lithospheric plates. Some of these sedimentary units are melted and intruded as igneous rocks; others are crumpled and raised into large mountain belts, where chemical attack by acidic water once again releases these ions to the sea. In effect, this is a grand sedimentary cycle. Elements bonded into solid minerals in rocks are put into solution and transported to the ocean by rivers, where organic and inorganic processes cause them to be precipitated into solids (such as halite, and silica and carbonate shells) and reincorporated into sedimentary rocks that become uplifted by tectonic processes and weathered once again, repeating this grand cycle.

If rivers are the primary source of salt ions to the oceans, why aren't the ionic compositions of freshwater and seawater similar? They clearly are not, as a comparison of Tables below shows. Shouldn't they be, if one is supplying material directly to the other? The difference in the relative composition of solutes in seawater and river water is a result of the residence time of ions in the ocean, which is simply the average length of time that an ion remains in solution there. It ranges between 2.6×10^8 years for sodium and 1.5×10^2 years for aluminum. This is no different from saying that your "residence time" in your bedroom, asleep in bed, is eight hours a day. The two most abundant components of seawater (Na^+ and Cl^-) have long residence times, on the order of hundreds of million years. Their persistence in a dissolved state in ocean water reflects their low geochemical and biochemical reactivity; in other words, they are essentially inert. By contrast, many of the principal ions in river water are characterized by short residence times in the oceans because they are much more reactive or are important for biological cycles. For example, many marine organisms require dissolved Ca^{2+} to secrete their carbonate ($CaCO_3$) shells. These calcium ions are in constant demand by the marine biota, so calcium has a relatively low residence time of 8×10^6 years.

Long residence times also help to explain the principle of constant proportions. Water is stirred and mixed by ocean currents, much as stirring a pot of soup with a spoon creates eddies and swirls (turbulence) that mix the ingredients. Studies of currents indicate that mixing rates in the oceans are on the order of a thousand (10^3) years or less. This rate is much lower than are the residence times of the major ions of seawater, which range from millions (10^6) to hundreds of millions (10^8) of years. Thus, rapid mixing and very long residence times of salt ions in seawater assure that these substances are distributed uniformly throughout the oceans. A more familiar way to think about this is to imagine yourself making a cake. Slowly adding dye along the edge of a bowl (this is equivalent to a river supplying ions at the edge of an ocean basin) to a cake batter that is being rapidly stirred by an electric beater (this is equivalent to mixing by ocean currents) quickly distributes the dye molecules evenly throughout the batter, so that its color (this is equivalent to the ocean's salinity) is uniform.

Table: Residence in Ocean waters.

Substance	Residence Time ($\times 10^6$ yr)
Chloride (Cl^-)	∞
Sodium (Na^+)	260
Lithium (Li^+)	20
Strontium (Sr^{2+})	19
Potassium (K^+)	11
Calcium (Ca^{2+})	8
Zinc (Zn^{2+})	0.18
Barium (Ba^{2+})	0.084
Cobalt (Co^{2+})	0.018
Chromium (Cr)	0.00035
Aluminum (Al)	0.00015

Effects of Salinity on the Properties of Water

Figure: Rapid mixing spreads dye evenly throughout the cake batter.

Here we learned a great deal about the structure of the water molecule and the salinity of the ocean. Now we are ready to examine the effect dissolved ions have on the physical properties of water. As you might guess, the addition of salt modifies some of the properties of water in a number of significant ways. Most of these changes come about because the ions are hydrated, which modifies the chemical behavior of the H_2O molecules in the solution. Let us explain by examining some specific water properties as they are affected by solutes.

Freezing Point

Pure freshwater freezes at 0°C. The addition of salt to the water lowers its freezing point. For example, seawater with a salinity of 35‰ freezes at a temperature of -1.91°C. The reason that the freezing temperature of seawater is depressed relative to that of freshwater is quite simple. The hydrated salt ions "hold on" to individual H_2O molecules, interfering with their rearrangement into an ordered ice crystal.

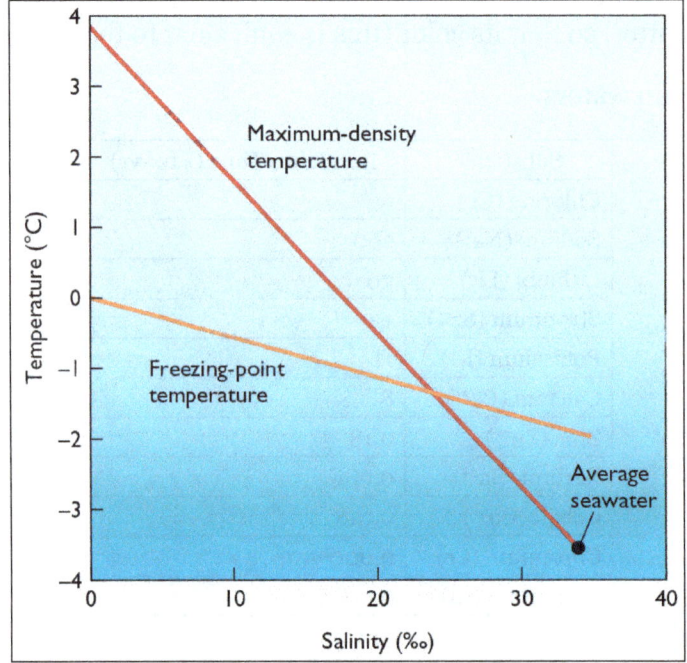

Figure: Effects of salinity on the maximum-density and freezing-point temperatures of seawater.

The addition of dissolved ions to water lowers the initial freezing point temperature of the solution because the hydrated ions interfere with the rearrangement of the H_2O molecules into an ordered ice structure. Also, an increase in salinity depresses the maximum-density temperature of seawater. Consequently, the temperature of maximum-density of average seawater is well below its freezing point. This means that seawater of average salinity will freeze before it will sink.

Density

Because solutes have a greater atomic mass than do H_2O molecules, the density of water increases with salinity. This means that freshwater floats on salt-water. For salinities >24.7‰, the temperature of maximum density (3.98°C for freshwater) is below the freezing point (0°C).

Vapor Pressure

The vapor pressure on a liquid surface is the pressure exerted by its own vapor. When a liquid such as water is placed in a closed container, some of the molecules will vaporize, decreasing the amount of liquid. Once equilibrium is reached, the vapor pressure is the pressure exerted by the molecules in the vapor phase. As the salinity of water increases, vapor pressure drops; this means that fresh-water evaporates at a faster rate than does seawater. The depression of vapor pressure in seawater is directly related to the total number of solute molecules. This effect is a consequence of the hydrated ions, which "hold on" to the water molecules, making their vaporization more difficult.

Chemical and Physical Structure of the Oceans

Oceanographers have acquired a great deal of information about variations in the fundamental properties of ocean water, including the regional distribution of temperature, salinity, and density along the sea surface and throughout the ocean's depths.

Temperature

Latitude exerts a strong control on the surface temperature of the ocean because the amount of insolation (the solar energy striking the Earth's surface) decreases poleward. Surface-water temperatures, therefore, are highest in the tropics and decrease with distance from the equator. Isotherms, imaginary contour lines that connect points of equal water temperature, generally trend east–west, parallel to the lines of latitude. Because the amount of insolation varies with the seasons, the surface-water temperature changes with time as well. The intense sunlight in the tropics and subtropics produces a broad band of water with a temperature higher than 25°C that shifts north or south depending on the season. Water in the polar oceans is very cold, much of it being near or even below 0°C. Seasonal shifts of the isotherms are minor in the ocean off Antarctica but are significant in the North Pacific and North Atlantic Oceans.

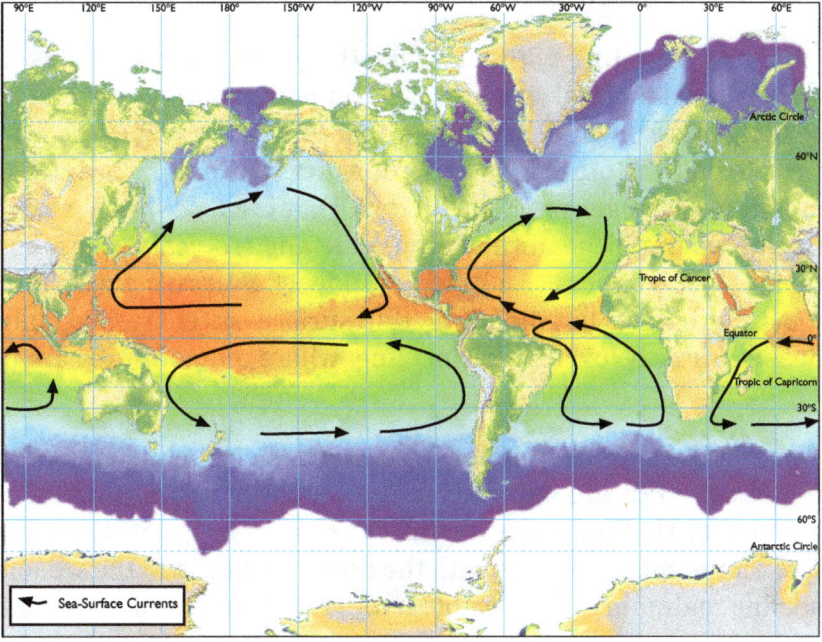

Figure: (a) Sea-surface temperature in August.

Ocean currents that flow around the periphery of each ocean affect the distribution of surface-water temperature. For example, the strips of >25°C water in the tropical Atlantic and Pacific Oceans are much broader at their western than at their eastern margins. This distortion is produced by currents that move warm water pole-ward along the western side of oceans and cold water equatorward along their eastern sides.

Because water is heated by the sun, and solar radiation decreases with distance from the equator, sea-surface temperatures vary directly with latitude. Deep blue indicates cold water (0–1°C) and orange, warm water (24°C+). Large-scale ocean circulation transports warm water poleward at the western edges of the ocean basins and cold water equatorward at their eastern peripheries. (a) Global sea-surface temperature (SST) is shown for August. (b) Global SST is shown for February.

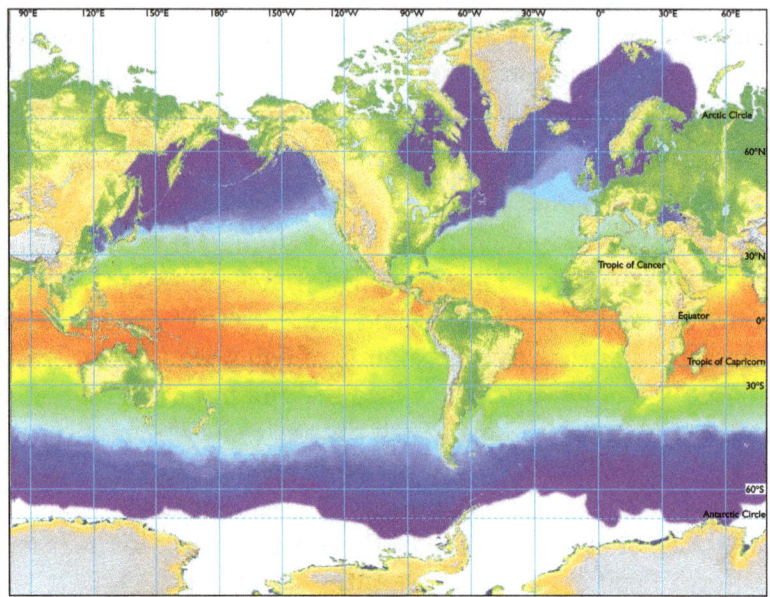

Figure: (b) Sea-surface temperatures in February.

Investigations of water temperature with water depth reveal that the oceans in the middle and lower latitudes have a layered thermal structure. A layer of warm water, several hundred to a thousand meters thick, floats over colder, denser water that fills up the rest of the basin. The two water masses are separated from each other by a band of water that has a sharp temperature gradient (meaning that temperature changes rapidly with depth); it is called the thermocline. The thermocline is a permanent hydrographic feature of temperate and tropical oceans, and ranges in water depth between 200 and 1,000 meters (~660 to 3,300 feet). Surprisingly, in terms of volume, the most typical water of the tropics is not the warm, thin surface-water layer so familiar to sailors and tourists, but the deeper water below the thermocline, where temperatures are near freezing even at the equator.

Because of solar heating during the summer, a seasonal thermocline appears at a depth of between 40 and 100 meters (~132 and 330 feet) in the oceans of the mid-latitudes. The daily thermoclines that occur at very shallow water depths (<12 meters; ~39 feet) are diurnal. Unlike the surface water, where temperature changes with the hour, the day, and the season, water below the permanent thermocline maintains a low, stable temperature over time, averaging <4°C for most of the ocean.

Salinity

The salinity of the ocean's surface water, although more complicated in detail than the pattern of the sea-surface temperature, also shows a clear latitudinal dependence.

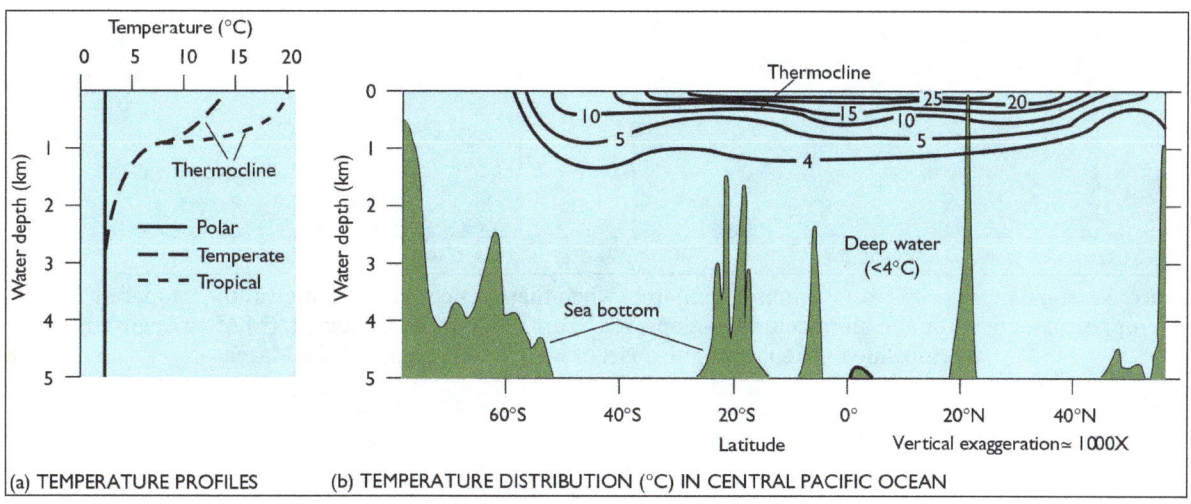

Figure shows temperature profiles. (a) These vertical profiles depict variations of temperature with water depth. Note the prominent thermocline that separates cold, deep water from warm surface water. (b) A longitudinal profile of the average temperature distribution in the Pacific Ocean using isotherms indicates that the bulk of all ocean water is colder than 4° C.

The highest salinity values occur between 20 and 30 degrees north and south latitude and decrease toward the equator and the poles. Salinity variations are caused by the addition or removal of H_2O molecules from seawater. Processes that remove H_2O molecules from seawater include evaporation and the formation of ice. Precipitation (rain, snow, and sleet), river runoff, and ice melting add H_2O molecules, diluting the salinity of seawater and reducing its density. Because these processes are dependent to a large degree on climate, and climate varies with latitude, the salinity of surface seawater varies directly with latitude.

Figure shows exactly how ocean salinity changes with latitude. Maximum salinity values occur in the subtropics and minimum values near the equator and the Polar regions. The dashed line in this figure is a plot of the latitudinal difference between evaporation and precipitation—that is, evaporation (removal of H_2O) minus precipitation (addition of H_2O). To calculate this value for any latitude, the total amount of evaporation for one year is subtracted from the total precipitation for that same year. If the result of the calculation is a positive value, then evaporation has exceeded precipitation, and salinity in this region will be higher than normal. A negative value denotes the converse, that is, precipitation exceeds evaporation, and salinity here is lower than normal. The rate of evaporation depends strongly, although not exclusively, on temperature and, thus varies directly with latitude, being very low in the polar and sub-polar regions, rising in the middle latitudes, and being highest in the tropics and subtropics. By contrast, rainfall maxima occur in the tropics and temperate latitudes, and minima in the subtropics and high latitudes. Therefore, during the course of a year, there is a net excess of precipitation over evaporation in the tropics and temperate latitudes, and an excess of evaporation over precipitation between 20 and 35 degrees latitude, where, not surprisingly, most of the major land deserts occur.

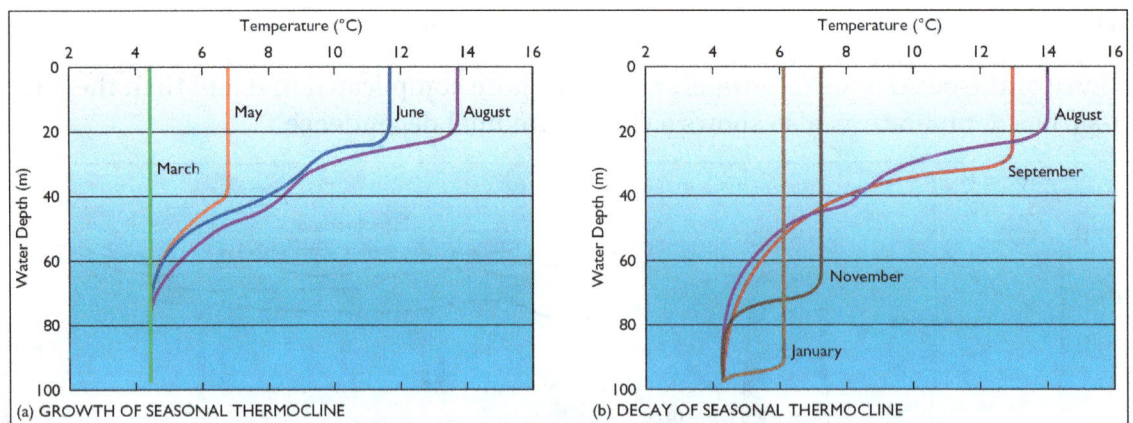

Figure: Seasonal thermocline. (a) Beginning in March, when there is no vertical change in the temperature of the upper water column, the thermocline develops; it is most pronounced in August. (b) After August, the thermocline weakens with the onset of winter and disappears by March.

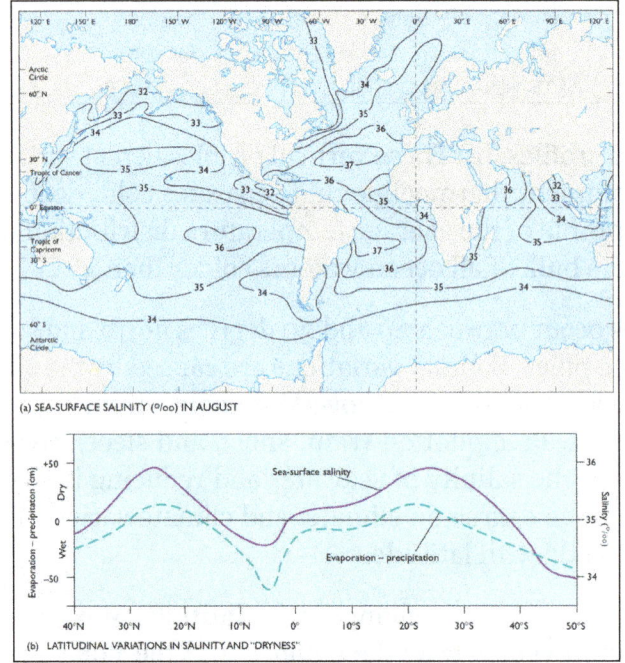

Figure: Salinity variations.

Figure shows (a) The surface salinity of the world's oceans in parts per thousand (‰) during August shows a regular pattern that depends on latitude, with maximum values in the center of each ocean and minimal values at the equator and Polar Regions. (b) A comparison of evaporation and precipitation rates accounts for the maximum salinity levels in the subtropics and the minimum salinity levels near the equator and the Polar Regions. This profile is a global average.

Notice in Figure above that the two curves representing evaporation minus precipitation and sea-surface salinity are quite similar in shape. This suggests that the major control on the surface salinity of the world's oceans is the relative effect of evaporation and precipitation—in other words, of climate. The maximum salinity levels in the subtropical oceans are produced by a strong excess of evaporation over precipitation. At the equator, evaporation rates are high, but rainfall is even greater, leading to the lower surface salinity in these waters. In the midlatitudes, rainfall rates surpass evaporation rates to an even greater degree than they do in the tropics, which reduces salinity to less than 34‰. Surface salinity in the polar seas fluctuates as ice forms and melts with the changing seasons. The water of the coasts and continental shelves is diluted by the freshwater of rivers. A case in point is the Amazon River, which injects more than 5×10^{12} cubic meters ($\sim 1.8 \times 10^{14}$ cubic feet) of fresh water into the western South Atlantic Ocean each year, lowering the salinity there.

As was the case for temperature, sharp salinity gradients characterize the water column of the world's oceans. These gradients are termed haloclines and, like temperature, represent boundary zones between distinct water masses. A north–south longitudinal profile of salinity in the western Atlantic Ocean reveals a well-developed layering of water masses. Water stratification (layering) is evident between 40°N and 40°S latitudes, where a lens of high-salinity (35‰) surface water is separated from less saline water below by a sharp halocline. Also, two tongues of water with salinities of <34.74‰ extend northward from the Antarctic region: one at about a 1,000-meter (~3,300-feet) depth, the other along the deep-sea bottom. These tongues indicate that their water originated at the sea surface in the south polar seas and are separated from each other by a water mass that flowed southward out of the North Atlantic Ocean. What is truly remarkable about this North Atlantic water below a depth of 2 kilometers (~1.2 miles) is its uniform salinity of ~34.9‰. This condition reflects the fact that once water sinks, it is no longer in contact with the atmosphere, where precipitation and evaporation alter the salinity. Thus, the salinity of deep water remains relatively stable (unchanging) over time, although very slow mixing processes, not well understood, do eventually change the salinity of water masses.

Density

Density, the amount of mass per unit volume (g/cm³), depends on the temperature, salinity, and pressure of seawater. Because water is essentially incompressible, pressure affects density only in the deepest parts of the ocean and can be ignored. It should, however, be noted that, if water was absolutely incompressible, sea level would be 50 meters (~165 feet) higher than it presently is! Water density controls the vertical structure of the water column, with the more dense water underlying the less dense water. Basically, the density of water is increased by dropping its temperature and raising its salinity. This means that cold, saline water is denser than is warm fresh water. The conversion of temperature and salinity data to water density can be done simply by consulting either tables or graphs.

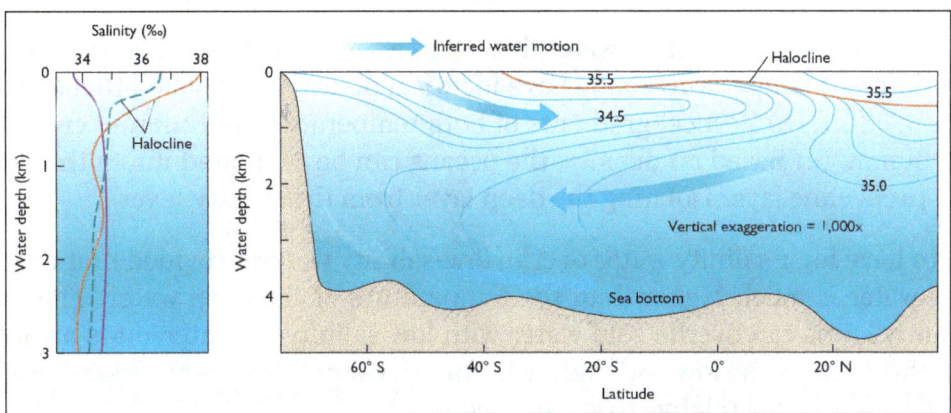

Figure: (a) Salinity profiles (b) Salinity distribution (%) in the western atlantic ocean.

(a) Vertical profiles of salinity show that sea-surface water may be more or less saline than the water below it. Note the prominent haloclines. (b) Isohalines (lines connecting points of equal salinity) in a longitudinal profile of the western Atlantic Ocean reveal distinct water-mass stratification and a prominent halocline. Below a 2-kilometer depth, the water has a remarkably uniform salinity, ranging between 34.7 and 34.9‰.

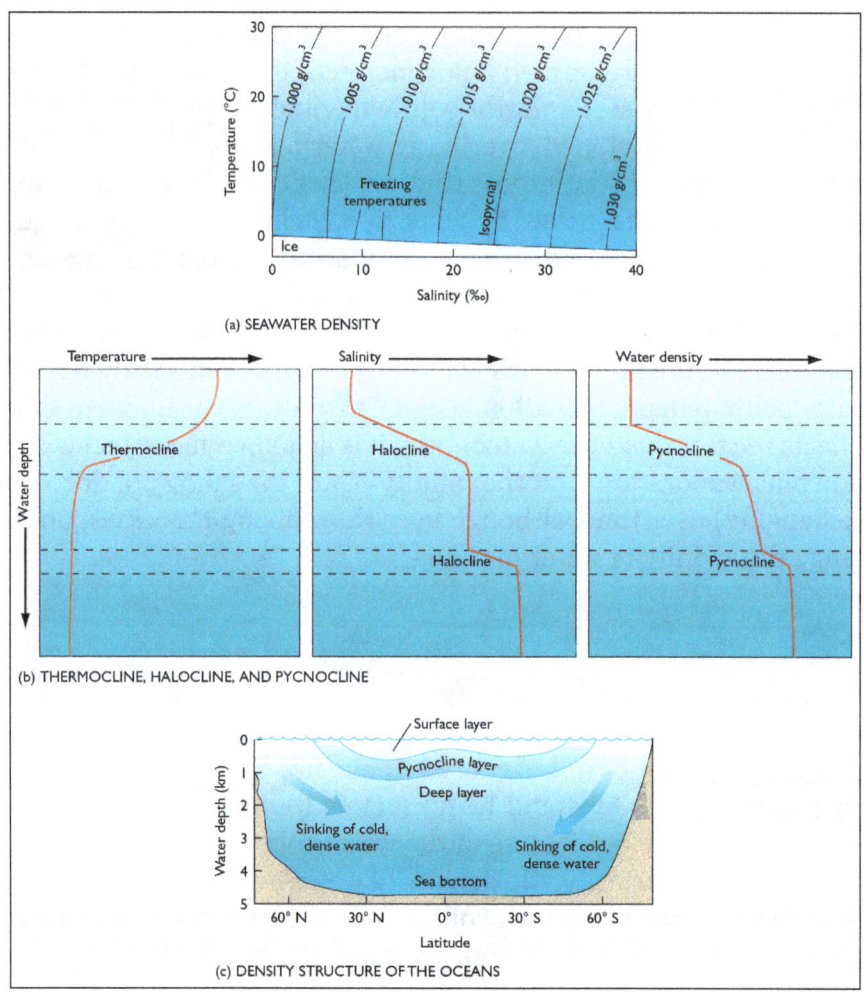

Figure: Density of seawater.

(a) This plot shows the variation of seawater density as a function of temperature and salinity. Water parcels of different temperatures may have identical densities, provided that salinity counters the temperature effect. (b) Vertical gradients of both temperature and salinity create pycnoclines in the water column. (c) Based on density, the oceans can be separated into a three-tiered structure, with the pycnocline layer isolating the deep layer from the surface layer.

It is possible to have high-salinity water overlie low-salinity water, provided that the temperature of the surface water is much higher than the temperature of the deep water. Thus, warm water with high salinity levels can overlie cold water with low salinity—a water-mass arrangement that is common in the ocean of the low and middle latitudes, where high evaporation rates increase the salt content of surface water relative to subsurface water.

If vertical gradients of water temperature (thermoclines) and salinity (haloclines) exist in the ocean, then it stands to reason that density will show a similar gradient because it depends directly on variations of temperature and salinity. A sharp density gradient with depth is called a pycnocline. Density stratification of the ocean imposes a three-layered structure on the water column: a surface layer, a pycnocline layer, and a deep layer.

Surface Layer

This topmost layer, the surface layer, is thin, averaging about 100 meters (~330 feet) in thickness, and represents about 2 percent of the ocean's volume. Being at the surface, it is the least dense water of the water column, largely because of its warm temperatures. Because it is in contact with the atmosphere, its water is affected by weather and climate, which cause diurnal, seasonal, and annual fluctuations of salinity and water temperature. Light penetrates to the bottom of this zone, allowing plant photosynthesis to occur wherever nutrient levels are adequate. In the Polar Regions, cooling of the surface water produces dense water that sinks. This sinking process prevents the formation of a pycnocline in the oceans of the high latitudes.

Pycnocline Layer

The boundary zone between the surface-water and deep-water layers are the pycnocline layer, where a sharp density gradient exists. It is not really a distinct water mass, but a transition zone where the surface-water layer grades into the deep-water layer. (Because temperature is the dominant control of water density in the pycnocline, many oceanographers commonly refer to this layer as the thermocline.) Water in this transition zone amounts to about 18 percent of the ocean's volume. In the low latitudes, the pycnocline corresponds to the permanent thermocline, which is created by the strong and persistent heating of the water by the tropical sun. In the midlatitudes, the pycnocline weakens and coincides with the halocline, which is produced by the abundant rainfall that dilutes the salinity of the water in the surface layer.

Deep Layer

About 80 percent of the total volume of the oceans is in the deep layer. The bulk of this deep water originates in the high latitudes, where it is cooled while in contact with the frigid polar atmosphere. This cold ($<4°C$) polar water sinks to the ocean bottom because of its high density, and flows slowly equatorward, supplying the depths of the ocean with water that was once near the ocean's surface. Gas concentrations depend a great deal on the density structure of the water column.

Gases in Seawater

To understand better the levels and distribution of gases in seawater, we need a thorough grasp of certain chemical concepts. The first is saturation value, which refers to the amount of gas at equilibrium that can be dissolved by a volume of water at a specific salinity, temperature, and pressure. The higher the saturation value is for a gas, the greater is its solubility, which is the property of being dissolved and going into solution. The solubility of gases increases with a drop in either water temperature or salinity and a rise in pressure. This means that cold, brackish (slightly salty) water can dissolve more gas than can warm, saline (salty) water at the same pressure. If a parcel of water is saturated with a gas, a change of water temperature or salinity will cause the quantity of gas to be below or above the water's new saturation value, conditions called under saturation and super-saturation, respectively. Water that is undersaturated can dissolve more gas, whereas water that is supersaturated can release gas from solution (bubbles may form under the right conditions).

At this point, it is necessary to specify the types of units that chemists use to characterize the quantity of dissolved gases in water. We will use a volume measure expressed as milliliters—that is, a thousandth (10-3) of a liter—of gas dissolved in 1 liter of water, expressed as ml gas/l. The amount of gas in a 1-milliliter volume depends on the temperature and pressure of the system. Therefore, the concentration of a gas dissolved in 1 liter of water is specified for a standard temperature of 0°C and a standard pressure of 1 atmosphere (equivalent to the pressure exerted by the weight of the atmosphere, which is the pressure that we feel at sea level).

The processes that produce, consume, and regulate gas concentrations in the ocean are summarized in table below and depicted in figure above. Enormous quantities of gases are exchanged between the ocean and the atmosphere near the sea surface. Ordinarily, surface seawater is near saturation with respect to the common atmospheric gases (O_2, N_2, and CO_2). However, gases diffuse across the air-sea interface, as the temperature and salinity of the water change or as organisms produce or consume gases, creating supersaturated or undersaturated conditions. Also, breaking waves, particularly those associated with large storms, drive air bubbles downward into the water, where water pressure dissolves some of the bubbles before they can ascend back to the surface. In fact, this very process—the passage of air bubbles through water—is used to aerate water in your fish aquarium. Oxygen diffuses out of air bubbles into the aquarium water where fish can then use it to "breathe."

Table: Summary of factors that regulate the concentration of gases in water.

Factors	Effects
Wave and current turbulence	Increases the exchange of seawater gases with the atmosphere.
Difference in gas concentration	Gases diffuse across the air-sea interface from high to low areas of concentration until chemical equilibrium is attained.
Temperature	A drop in water temperature increases the solubility of gases.
Salinity	A rise in salinity decreases the solubility of gases.
Pressure	A rise in pressure increases the solubility of gases.
Photosynthesis	Increases concentration of O_2; decreases concentration of CO_2.
Respiration	Increases concentration of CO_2; decreases concentration of O_2.
Decomposition	Increases concentration of CO_2; decreases concentration of O_2.
pH	Controls the relative concentrations of the various species of CO2 in water (H_2CO_3, HCO_3^-, CO_3^{2-}).

The primary regulator of gas concentrations in subsurface water is the activity of organisms. When light and nutrient conditions are adequate, plants photosynthesize. This complex biochemical process of photosynthesis converts water (H_2O) and carbon dioxide (CO_2) into organic matter and liberates oxygen (O_2) as a product of the reactions. Thus, plants simultaneously reduce the dissolved content of CO_2 and augment the levels of dissolved O_2 as they conduct photosynthesis in the upper, sunlit part of the water column during daylight hours. In contrast, respiration, the chemical breakdown of food in cells for the release of energy, is conducted by both plants and animals in surface water, and by animals at all depths of the ocean. This results in the uptake of O_2 and the release of CO_2 as organisms oxidize food for nutritional purposes. As you sit reading this passage, you are respiring, "burning" food in the presence of oxygen, which releases the energy that your body requires to live. Consequently, these life-sustaining processes— photosynthesis and respiration—have a profound impact on the concentrations of dissolved CO_2 and O_2 in the oceans. Last, dead organic matter and excrement are decomposed by microbes, chiefly bacteria. Enzymes that are secreted by microbes chemically "attack" organic matter and break it down into simpler chemical compounds by the process of oxidation, a chemical reaction that consumes O_2 and produces CO_2 and other gases.

Gases other than oxygen and carbon dioxide are also released and taken up by physical and biological processes. For example, the decay of unstable radioactive elements contained in the minerals of deep-sea muds produces a variety of gases, including helium (He), radon (Rn), and argon (Ar). Some gases, such as carbon dioxide and helium, are spewed out of active submarine volcanoes on spreading mid-ocean ridges. These gases are then slowly transported hundreds, perhaps thousands, of kilometers by deep-sea currents before mixing dilute them to such a degree that they are no longer distinguishable in the water column. Such gases can be used to trace the flow path of currents near the sea bottom, even very sluggish ones. Also, ocean basins filled with anoxic water (water without dissolved oxygen) are not populated by normal communities of plants and animals because these organisms cannot survive without O_2. However, specialized anaerobic bacteria, which live without free oxygen, inhabit such waters. They use the oxygen atoms that are chemically bonded to sulfate ions (SO_4^{2-}) for conducting their metabolic processes, which yields the toxic gas hydrogen sulfide (H_2S) as a reaction byproduct. Although anaerobic conditions rarely develop in the well-mixed, and thus well-ventilated, water of the open ocean, they do occur in some restricted basins where circulation is sluggish and the supply of oxygen is limited. For example, water trapped in the small basins that are part of the continental shelf of southern California commonly is anoxic because of sluggish bottom currents and poor mixing.

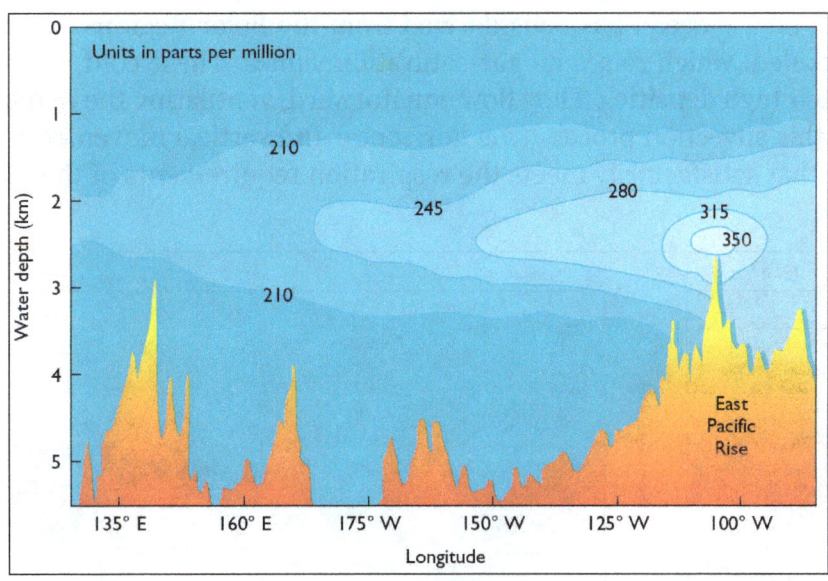

Figure: Helium-3. A large plume of water enriched in helium-3 extends westward from the East Pacific Rise, tracing the flow of water at mid-depth in the Pacific Ocean. Volcanic emissions are clearly the source of this gas, which originated deep in the Earth's mantle.

In order to firm up the gas chemistry of the oceans, sinks, and distribution of O_2 and CO_2—two gases vital for the ocean's biota. Both are considered to be non-conservative substances in seawater because their concentrations vary greatly over short time and distance scales.

Oxygen

The vertical distribution of O_2 in the ocean of the low and middle latitudes shows a distinct pattern. A warm surface-water layer with a high content of dissolved O_2 is separated from cold, relatively well-aerated deep water by a distinct oxygen-minimum layer at about 150 to 1,500 meters (~495 to 4,950 feet) below the sea surface. This vertical oxygen profile reflects inputs and outputs of gases by a variety of processes.

There are two principal sources of O_2 for the oceans—gas diffusion across the air-sea interface and plant photosynthesis. Both of these processes are limited to the uppermost levels of the water column and account for the high concentrations of dissolved O_2, typically in excess of 5 milliliters per liter, in the surface-water layer. The O_2-minimum zone coincides with the pycnocline layer; this implies a connection between the two. Organisms of all kinds are drawn to the pycnocline layer because of the ample food supply, and, as they respire, they deplete the O_2 content of the water. Also, bacteria there decompose the abundant dead organic matter, which further reduces the dissolved O_2. The sharp pycnocline indicates that the water column is stable and that vertical mixing of water is minimal at these intermediate water depths. Thus, the high demand for dissolved O_2 by animals and bacteria that are feeding on the dead and living organic matter, combined with a relatively slow rate of water mixing and thus of O_2 replenishment, produces the O_2-minimum layer with O_2 concentrations of <2 milliliters per liter.

Below the pycnocline, water is sparsely populated because food is scarce. The biological demand for dissolved O_2 is low here, and O_2 concentrations rise with depth to between 3 and 5 milliliters per liter. Because oxygen production is restricted to surface water, the dissolved O_2 found in this deep layer must have been derived from shallow depths. Temperature and salinity data indicate that deep water in all the ocean basins is derived from the Polar Regions. In the high latitudes, surface water is cooled, which raises its gas-saturation values. These cold, O_2-rich water masses sink because of their high densities. They flow equatorward, ventilating the depths of all the ocean basins. Although this advection process (the horizontal and vertical movement of a fluid) is slow, it occurs at a rate that satisfactorily meets the respiration requirements of the scanty populations of deep-sea fauna.

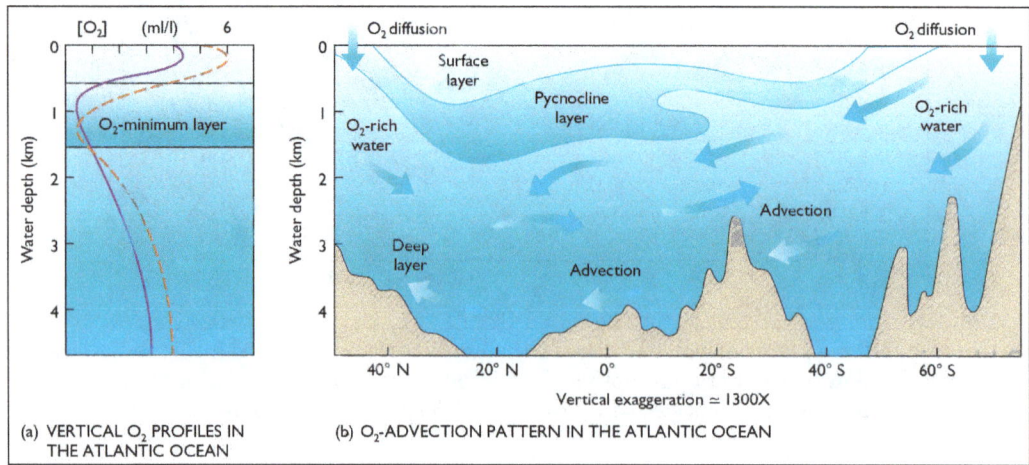

Figure: Dissolved oxygen. (a) Vertical profiles of dissolved oxygen reveal a distinct pattern consisting of well-ventilated surface and deep waters that are separated by an O_2-minimum zone in the pycnocline layer. (b) The deep water of the ocean is aerated by the advection of cold, dense, O_2-rich polar water.

Carbon Dioxide

Carbon dioxide (CO_2) is a gas that is actively involved in photosynthesis and respiration. Plants cannot produce food and survive in its absence. Animals release CO_2 as they break down food for energy by the chemical process of respiration. Also, CO_2 regulates the acidity of seawater. Let's examine this latter chemical property of seawater.

The degree to which water provides a suitable habitat for marine biota is determined, in part, by the concentration of dissolved hydrogen ions (H^+), a measure termed the pH of water. In pure water (water containing nothing but H_2O molecules) at 25°C, a very tiny fraction of the H_2O molecules, about 10^{-7}, or one in 10 million (10^7), spontaneously dissociates (breaks apart) into a hydroxyl (OH^-) and a hydrogen (H^+) ion. The free hydrogen ions are what control the acidity of water. The formal definition of pH is,

$$pH = -\log_{10}(H^+).$$

At first glance, this formula looks quite difficult to grasp, but it need not be. Remember that the pH of water is, to a first approximation, merely a measure of the concentration of the H^+ ion. The parentheses in the preceding equation can be equated to the concentration of H^+, although technically they refer to the chemical activity of this ion, which is an indirect function of its concentration level. Now, if we go back to our example of pure water, measurements indicate that 10^{-7} H_2O molecules (one in 10 million) are separated into their ionic components. If we substitute 10^{-7}, the concentration of H^+, into the formula and solve for its negative log, we get a pH value of 7. The log to the base ten of 10^{-7} is the exponent -7; thus, the negative log of 10^{-7} is -(-7) or 7. Water with a pH of 7 is, by definition, a neutral solution consisting of equal parts (10^{-7}) of OH^- and H^+. Raising the amount of H^+ to 10^{-6}—remember that 10-7, one H^+ ion in 10 million of the H_2O molecules, is a much smaller value than 10^{-6}, which is one H^+ ion in a million H_2O molecules—lowers the pH to 6, a solution that is no longer neutral but acidic. The log to the base 10 of 10-6 is the exponent -6; thus, the negative log of 10^{-6} is -(-6) or 6. Low concentrations of H^+ that impart a higher pH than the neutral level of 7 create a basic solution, in which the amount of OH^- surpasses the amount of H^+.

It's important that you understand the preceding discussion. A neutral solution has a pH of 7, which means that the H^+ concentration is 10^{-7}. A basic solution has a pH value that is greater than 7 (remember that this means low concentrations of H^+) and an acidic solution has a pH value that is lower than 7 (high H^+ concentrations). So the pH scale, to a first approximation, is nothing more than a measure of the content of H^+ in the water. Because the H^+ ion is so reactive with other compounds, acid water (water with a pH of less than 7 and thus a relatively high concentration of H^+), is a powerful dissolution agent, able to weather rocks chemically. Also, plants and animals living in the ocean are affected by the content of H^+ in the water they inhabit because many metabolic activities are regulated by the seawater's pH.

The pH of water is directly linked to the CO_2 system. When CO_2 is added to water, most of it is rapidly converted into carbonic acid (H_2CO_3) as it bonds with water molecules (hydration). This acid then rapidly dissociates into bicarbonate (HCO_3^-) and carbonate (CO_3^{2-}) ions, which yields H^+ ions, making the water acidic. The specifics of the chemical reaction are:

$$CO_2 + H_2O \rightarrow H_2CO_3 \rightarrow HCO_3^- + H^+ \rightarrow CO_3^{2-} + 2H^+.$$

pH	[H$^+$]	[OH$^-$]	
	1	10^{-1}	10^{-13}	Hydrochloric acid
	2	10^{-2}	10^{-12}	Lime juice
Acidic Solutions	3	10^{-3}	10^{-11}	Acetic acid
	4	10^{-4}	10^{-10}	Tomato juice
	5	10^{-5}	10^{-9}	Black coffee
	6	10^{-6}	10^{-8}	Milk
Neutral	7	10^{-7}	10^{-7}	Pure water
	8	10^{-8}	10^{-6}	Seawater
	9	10^{-9}	10^{-5}	Borax solution
	10	10^{-10}	10^{-4}	
Basic Solutions	11	10^{-11}	10^{-3}	Milk of magnesia
	12	10^{-12}	10^{-2}	Household ammonia
	13	10^{-13}	10^{-1}	Lye
	14	10^{-14}	10^{-0}	Sodium hydroxide

Figure: The pH scale. The concentration of H$^+$ in water is specified by pH. A pH of 7 denotes that the H$^+$ concentration is 10^{-7}, or one part in 10 million. Solutions with low pH values (high concentrations of H$^+$) are acidic. Solutions with high pH values (low concentrations of H$^+$) are basic.

This formula summarizes each step of the reaction when CO_2 is dissolved in water. Notice that the free ions of H$^+$ lower the pH of the water. Therefore, we conclude that the addition of dissolved CO_2 tends to lower the pH of water. At equilibrium, the proportion of each of the carbon compounds (H_2CO_3, HCO_3^-, and CO_3^{2-}) depends on the pH of the water. At the pH of normal seawater (7.8 to 8.2), about 88.9 percent of the carbon compounds occur as HCO_3^-. A change in the amount of any of these carbon substances disrupts the balance of the CO_2 system, and reactions occur to reestablish equilibrium. For example, the removal of CO_2 from water by plant photosynthesis or by solar heating, which reduces its saturation value and causes bubbles to form, initiates a series of chemical responses that shift the reactions specified in the formula to the left. This results in the production of CO_2. Respiration, in contrast, releases CO_2 into the water. This process causes the reactions in the formula to move to the right, which increases the concentrations of the other chemical species at the expense of CO_2.

Organisms are sensitive to pH, quite a few extremely so. Fortunately, the many rapid chemical reactions within the CO_2 system prevent large fluctuations in the pH levels of the world's oceans. Seawater is essentially a stable solution with a pH that rarely ranges below 7.5 or above 8.5. This condition is described as buffered, meaning that the mixture of compounds and the nature of the reactions are such that the pH of the solution is hardly affected despite an input or output of H+. Let me explain. Increasing the level of H$^+$, which should lower the pH (make the water more acidic), causes the reactions to shift to the left; as a consequence, H$^+$ is removed as it complexes with HCO_3^- to form H_2CO_3, buffering the solution (keeping the pH near its original value). Conversely, reducing the level of H$^+$ (making the solution more basic by raising the pH) reverses the reactions. There is a rapid release of H$^+$ into solution, as H_2CO_3 dissociates to HCO_3^- and releases H$^+$. The end result is that the pH is stable despite changes in the relative amounts of the carbon species in the CO_2 system; the system is basically self-regulating or buffered.

With this background, we can now explain the dissolution of $CaCO_3$ shells in cold, deep water but not in warm, shallow water, and the ocean's carbonate compensation depth (CCD). Recall that the pH level of water is inversely proportional to the concentration of dissolved CO_2. This means that the higher the CO_2 content of the water is, the lower the pH (or, if you prefer, the greater the content of H⁺). Cold water has a higher gas-saturation value than does warm water because of its low temperature. Also, the saturation value of gases in water increases with increasing pressure and therefore water depth. This means simply that the cold, dense water that fills the ocean depths contains high levels of dissolved gases like O_2 and CO_2. More importantly, respiration in the deep layer of the ocean releases CO_2 into the water. The high CO_2 concentration of the deep water lowers the pH, making the water acidic and dissolving $CaCO_3$ shells that sink to the deep-sea floor. Shallow water, on the other hand, is relatively warm and has a lower concentration of CO_2 at saturation than does the deep, cold water. This raises the water's pH level, which releases carbonate ions (CO_3^{2-}) that chemically bond to the abundant calcium ions (Ca^{2+}) and precipitate $CaCO_3$. Therefore, carbonate oozes tend to accumulate in relatively shallow water of the deep sea above the CCD because water below the CCD is cold and acidic and dissolves shells composed of $CaCO_3$.

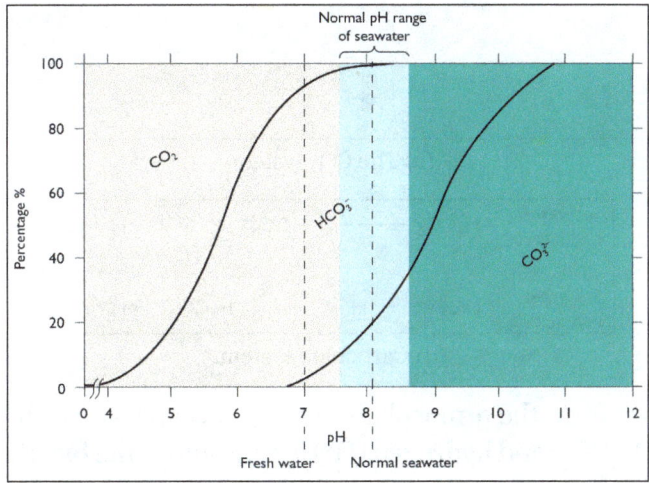

(a) Distribution of Carbon species in water.

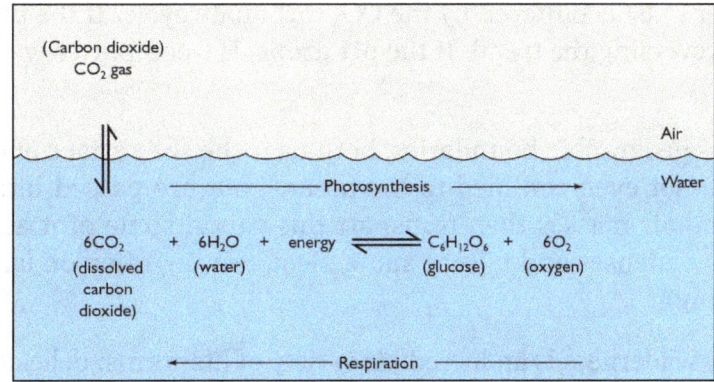

(b) Photosynthesis and respiration.

(a) Carbon dissolved in water occurs in the form of carbon dioxide (CO_2), bicarbonate (HCO_3^-), and carbonate (CO_3^{2-}). At equilibrium, the relative proportion of these carbon species depends on the pH of the water. Basic solutions are dominated by CO_3^{2-}; acidic solutions, by CO_2. At the pH of normal seawater, HCO_3^- makes up about 80 percent of the carbon species.

(b) The CO_2 system is also involved in biological processes, notably photosynthesis, which removes dissolved CO_2 from the water, and respiration, which liberates CO_2 into the water. Thus, the CO^{2-}-carbonate cycle influences and is influenced by both chemical and life processes.

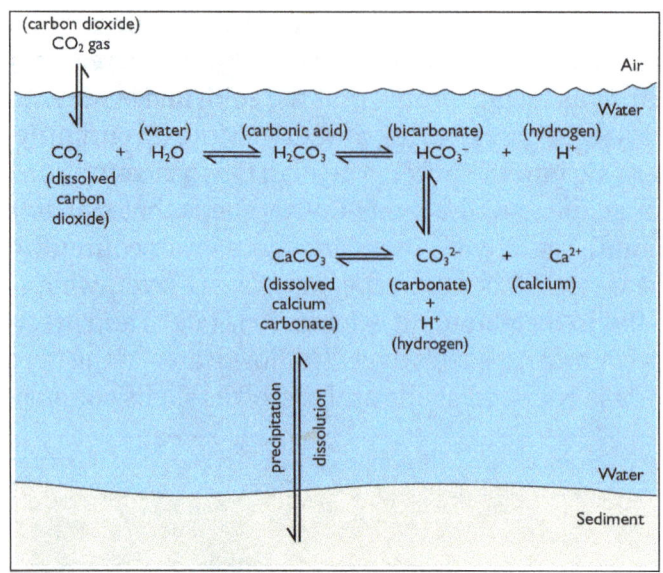

(c) The CO_2 system.

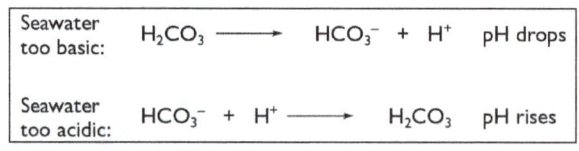

(d) Carbonate system.

(c) Carbon dioxide diffuses from the atmosphere into the oceans, where it complexes with water to form weak carbonic acid (H_2CO_3) and hydrogen (H^+) ions. Some of the bicarbonate ions dissociate into carbonate (CO_3^{2-}) ions, which may complex with calcium (Ca^{2+}) to form calcium carbonate ($CaCO_3$).

(d) The pH of seawater (7.8) is buffered by the CO_2-carbonate cycle. If the pH rises, H_2CO_3 dissociates and yields H+, reversing the trend. If the pH drops, H+ complexes with HCO_3^-, causing the pH to rise again.

Water does flow across geographic boundaries, both near the sea surface and at depth. H_2O molecules in seawater are not even confined to the oceans, but are passed into the atmosphere by evaporation; winds and air masses then transport this vapor, some of it across great distances. Eventually, the vapor condenses and falls as snow, sleet, or rain, often on land, whence it returns to the ocean as river runoff.

Dwelling in this watery wilderness is an incredible variety of life-forms: delicate, lacy plants; spindly legged shrimp; swift, powerful sharks and tuna; microscopic, floating algae; and solidly anchored oysters. We marvel at the ability of these creatures to live in the ocean, but many of us fail to realize that we, as terrestrial inhabitants, live in an ocean as well. We dwell on land beneath an ocean of air that is an order of magnitude (differs by a factor of 10) deeper than the deepest seas and that, because of its weight, exerts considerable pressure on us. For those of you who are technically inclined, atmospheric pressure at sea level is about 1,013 millibars, or ~1,013× 10^6 dynes/cm², or ~10 metric tons/m².

This transparent envelope of fluid—the gaseous atmosphere—is in direct contact with the surface of all the land and oceans and is a crucial element of the planet's water cycle. The atmosphere and hydrosphere, which enclose the Earth's crust, interact through a vast network of processes.

Powered by the energy of the sun, water is exchanged among the oceans, the atmosphere, and the landmasses through evaporation, precipitation, river flow, groundwater (subterranean water) percolation, ocean circulation, and a host of related and intertwined processes. Some of these operate quickly; others, quite slowly. However, before we examine the workings of this immense global water cycle, we need to consider the distribution of water on the Earth.

Reservoirs of Water

Water covers more than 60 percent of the Earth's surface in the Northern Hemisphere and over 80 percent in the Southern Hemisphere. Not surprisingly, most of the water on the Earth—in fact, 97.25 percent of it by volume—is found in the oceans. It may astonish some of you to discover that rivers and lakes, which are so familiar and so indispensable to the activities of humans, are of little significance in the Earth's overall water inventory. In fact, groundwater—the water below the ground surface that saturates the void space in soil, sediment, and rock—exceeds the combined volume of water in lakes, rivers, and the atmosphere by one order of magnitude. Also, the atmosphere contains a significant quantity of water, both as gas and as droplets (clouds), that far exceeds the volume of water stored in rivers. Finally, a minuscule quantity of water is stored in the biosphere, in the cells and tissues of plants and animals.

Table: Earth's water reservoirs.

Reservoir	Water Volume (10^6 km³)	Total Water (%)
Oceans	1370	97.25
Ice masses	29	2.05
Groundwater	9.5	0.68
Lakes	0.125	0.01
Atmosphere	0.013	0.001
Rivers	0.0017	0.0001
Biosphere	0.0006	0.00004
Total	1408.64	99.99

Although the values cited in Table above are accurate for the present, they do not reflect the quantities that existed in some of these reservoirs in the geologic past. During glacial ages, for example, seawater that evaporated from the oceans fell as snow on land and was converted into ice, forming enormous ice caps and mountain glaciers. So, at that time, the volume of water stored in the ice reservoir was much greater than it is at present. Obviously, water does not reside indefinitely in any one state or reservoir; rather, it is continually changing back and forth, from gas to liquid to solid, and shifting from one reservoir to another. At present, global warming is causing a worldwide melting of mountain glaciers and ice sheets.

The Global Water Cycle

The exchange of water among the ocean, the land, and the atmosphere is termed the hydrologic cycle. The sun's heat evaporates surface water from the ocean and the land. Evaporated water enters the atmosphere as vapor and most of this return directly to the sea as precipitation. Air currents transport the remainder of this water vapor over land, where it condenses and falls as rain or snow.

This precipitation flows as runoff into rivers, collects temporarily in lakes, ponds, and wetlands, infiltrates the ground (groundwater) only to emerge later in rivers and lakes, or remains in solid form as snow or ice. With time, all of this water finds its way back to the oceans, either as river outflow, melting glaciers and icebergs, groundwater seepage, or evaporation and precipitation, thereby closing the hydrologic cycle. In effect, the same molecules of water are being continually recycled from reservoir to reservoir, the rates of flux and the quantity exchanged and stored depending on climate.

Over the oceans, evaporation exceeds precipitation, and the balance of water is maintained by river inflow. On all the land combined, in contrast, precipitation far exceeds evaporation, and the excess water travels to the ocean by river and groundwater flow. Estimates of these various fluxes are summarized in table below. These calculations assume that the total volume of water on the planet is fixed on a global scale (a perfectly reasonable supposition on a short time scale), so that precipitation and evaporation are balanced for the planet as a whole.

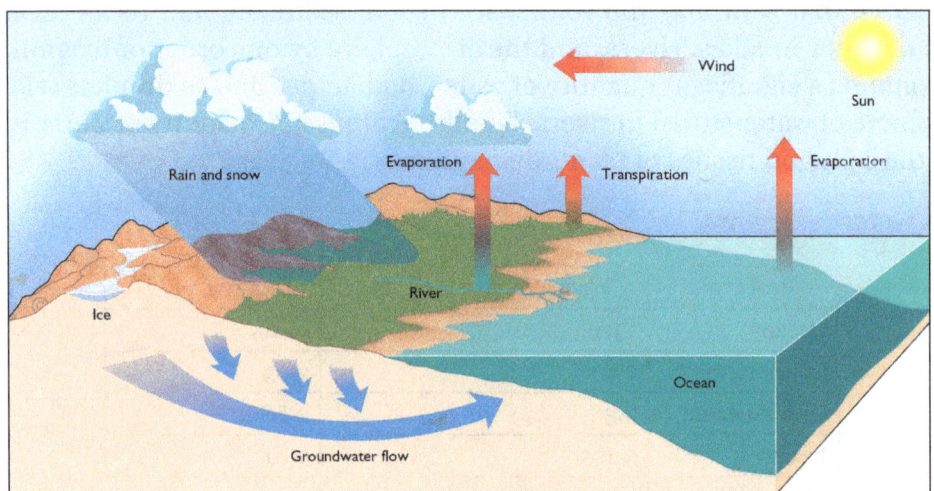

Figure: Hydrologic cycle.

The hydrologic cycle incorporates the major pathways for the transport and exchange of water among the various reservoirs of the earth. Basically, water circulates and changes phases (solid, liquid, gas) continually. In the gaseous state, water is supplied to the atmosphere by the evaporation of surface liquid water, much of it from the oceans, and by transpiration, the passage of water vapor through the surface of leaves. Condensation then causes the atmosphere to release this water as a liquid (rain) and as a solid (snow). These liquid and solid forms of water eventually return to the oceans by river runoff and groundwater flowage, closing the hydrologic cycle.

Figure: Recycling of water. Water evaporated from the ocean condenses as clouds that drop rain onto these mountains. River runoff and groundwater flow discharge this water into the ocean, closing the hydrologic cycle.

Table: Water fluxes.

Process	Water Flux (km³/yr)
Evaporation from land	72,900.
Precipitation on land	110,300.
Precipitation on oceans	385,700.
Evaporation from oceans	423,100.
Total precipitation on Earth	496,000.
Total evaporation on Earth	496,000.

The Ocean as a Biogeochemical System

Now that we have examined the chemistry of seawater and the physical nature of the water column, we can combine them into a generalized biogeochemical system. Rivers supply the ocean with most of its dissolved ions, these being derived mainly from the chemical weathering of rocks on land. The driving force underlying the ocean's biogeochemical system is the sun. Solar radiation absorbed by ocean water raises its temperature and stratifies the water column. A thermocline, separating the warm surface water from the cold deep water, forms.

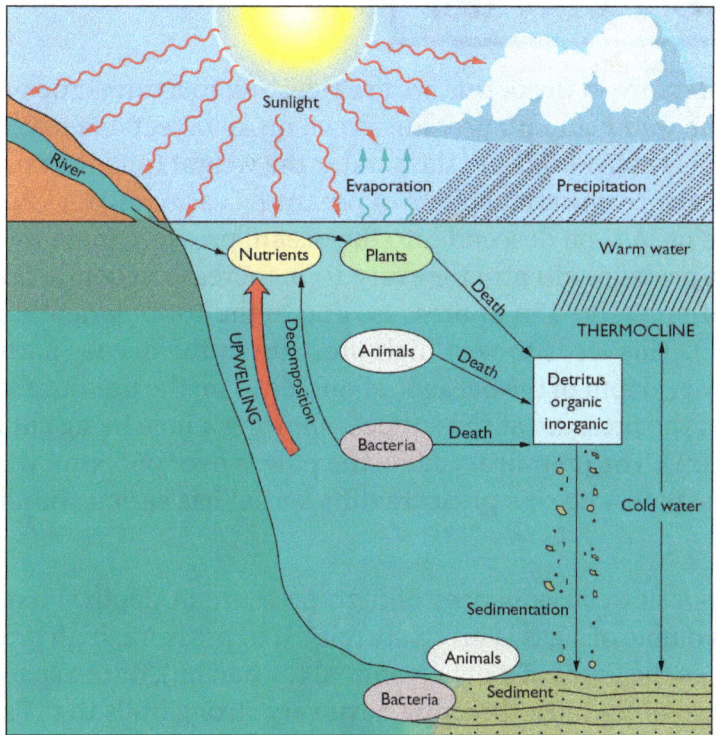

Figure: Biochemical recycling of matter.

Inorganic nutrients are converted into food by plant photosynthesis in the surface-water layer of the ocean. Animals eat plants and one another. When plants and animals die, their organic matter settles through the water column, where it is converted into simple nutrients by bacterial decomposition. This nutrient-charged water is then returned slowly to the surface by upwelling currents, completing the biochemical recycling of key nutrients.

In addition, sunlight is used by plants to convert carbon dioxide and nutrients into organic compounds (food) by photosynthesis. Small marine animals that graze on the multiplying plants in turn serve as food for even larger animals. This results in the transfer of energy (radiation from the sun) and matter (nutrients from rocks) through biological systems. When marine organisms die, they sink below the permanent thermocline, taking with them nutrients (phosphorus and nitrogen compounds) that are so crucial to plant photosynthesis. Bacteria decompose this dead organic matter, freeing these substances and producing nutrient-rich water. The vertical mixing of ocean water then slowly transports this water and its dissolved load of chemicals upward, re-infusing the sunlit surface layer with nutrients that, once again, are incorporated into living plant tissue by photosynthesis. Some of these materials collect on the sea bottom as oozes, where they become buried and lithified into rock, and by the process of plate subduction, are deformed and raised to form a mountain belt. As these rocks are squeezed and uplifted by the tremendous forces of the colliding plates, they undergo chemical weathering and mechanical erosion until the chemicals are released once more and transported by rivers to the ocean. The result is a continuous biochemical recycling of elements between living and nonliving matter and between surface and subsurface water.

Pressure in Ocean Water

When we talk about pressure in the ocean, we are referring to hydrostatic pressure, which is a result of the weight of the water column pressing down on an object due to gravity. The deeper you go, the more water that is above you, and the greater the weight (and thus pressure) of that water. At the surface we experience one atmosphere of pressure (1 atm = 101.3 kPa) due to the weight of the atmosphere above us. As you descend into the ocean, pressure increases linearly with depth; there is an increase in pressure of 1 atm for every 10 m increase in depth. So at 1000 m depth the pressure would be 101 atm (100 atm of pressure due to the 1000 m depth, plus the 1 atm that is present at the surface). This leads to very high pressure in the deeper parts of the ocean; if you consider that the average depth of the ocean is about 3800 m, the pressure at that depth would be 381 times greater than the pressure at the surface. showing a hole being cut into a pipe at a depth of 6 miles (about 9.6 km). The pressure outside the pipe is over 960 atm, which is far higher than the pressure inside the pipe, and this pressure differential has serious implications for a passing crab.

There are several important consequences of high pressure at depth. First, due to Boyle's Law, which states that the volume of a gas is inversely related to pressure, high pressure will act to compress air spaces, such as the lungs of a diving animal (or person), or the space inside a submarine. Submarines and submersibles must therefore have very strong hulls to resist this compression at extreme depths. Second, Henry's Law provides that at higher pressures a fluid will contain more dissolved gas. This means that deeper, high pressure water may contain more dissolved gases than surface water.

This also has implications for human divers. If you increase pressure, you increase the amount of gas that can dissolve in a fluid (such as blood). Conversely, when you reduce pressure, the fluid holds less dissolved gas, and the excess gas will leave the solution, often in the form of bubbles. This is what happens when you open a bottle of a carbonated beverage. The contents in the bottle are sealed under pressure, and as you open the bottle, you release the pressure, and the fluid can no longer hold all of the CO_2 that was dissolved in it, so the CO_2 escapes, forming bubbles. Decompression sickness or "the bends" occurs in SCUBA divers if they ascend too quickly after breathing compressed air. A slow ascent allows this excess gas to be removed from the blood and exhaled, but if the diver ascends too quickly, these gases will come out of solution and form bubbles in the blood that congregate near the joints, causing intense pain and perhaps death.

Grand Challenges in Physical Oceanography

Physical oceanography is a relatively young scientific branch, in comparison with marine biology or other oceanographic disciplines, and with most physics disciplines. However, because of importance for fisheries and trading, surface ocean physics on currents and waves has been investigated for over centuries. It was largely applied science; a lot of fundamental science on the ocean interior is probably still ahead of us. This may have to do with the ocean's inaccessibility for mankind.

Present-day Problems

The physical environment and variations in its governing processes have direct impact on sea-life, economics, atmospheric, and climate science, fluid dynamics issues and all general oceanographic disciplines. This may not be considered as a problem, rather as an exciting challenge, except that our present-day knowledge of, especially deep-ocean, processes is rather shallow. Physical oceanographers have difficulty in putting numbers to the vertical flux of material (e.g., nutrients) into the photic zone; they cannot precisely predict the effects of varying climate on ocean circulation and sediment transport; they do not know whether hurricanes destroy living organisms and their ecology in the deep ocean.

Part of the problem is a lack of dedicated observations, to such extent that the phrase "We know more about the moon than the (deep) ocean" (e.g., Thar) is still very much accurate. The difficulty lies in money spent, in coping with the rather harsh conditions of ambient pressure that increases by 1 Bar (one "atmosphere") every 10 m, of the salt content and, in shallow waters, of sedimentological erosion and biological growth deteriorating instrumentation, while, in deeper waters, lack of light does not allow a single sun's photon to reach below 1,000 m. Making ocean observations is not straightforward, also because of the vastness of the ocean requiring careful planning before going out to sea, of the working from unstable platforms like ships riding surface waves, while modern electronics do not endure a single drop of sea water on their circuit boards. Satellite observations tell us a lot about the ocean's near-surface, but not much about the ocean interior. The ocean is far more transparent for acoustics than for light, but acoustics yield limited quantitative research on ocean dynamics as their reflectors vary in shape and density. Limited resources force the oceanographer to make observations in spatially one-dimensional 1D profiling, whether from a ship or using self-contained instrumentation moored between a buoy and an anchor at the bottom.

It is thus not surprising that a substantial amount of research on geophysical flows, in particular on ocean dynamics driving the large- and small-scale circulation, has been performed in theory mainly, so far. In doing so, the limitations start with the presently still un-solvability (analytically) of the primary Navier-Stokes "NS" equations, which themselves form a rather comprehensive set. Although a good number of reasonable approximations are made for particular flows, the inherent nonlinearity of these equations is one of the outstanding problems in physics. It leads to perhaps the biggest challenge in classical physics, according to Einstein and very much valid to date, implying that no mathematical theory exists from first principles that describe nonlinear dynamics leading to turbulent flows that are omnipresent in geophysical flows. Without turbulence there is no marine life, no transport of matter, no ocean dynamics.

But, if a mathematical solution is in-existent one could nowadays resort to numerical simulations that, given the rapid development of modern electronics, may find a "reasonably accurate" approximate solution. Unfortunately, like clouds forming a big problem in (atmospheric) climate modeling and science, we are still far from properly incorporating small-scale processes in the large-scale ocean circulation and even in regional models. In spite of tremendous improvements in computing skills over the last decades, we do not have a clue whether one wag of an anchovy's tail will alter the Kurushio Current or generate an underwater storm.

Apart from being able to understand certain individual processes, we have difficulty understanding more complex coupled processes. For example, the ocean may become warmer, but that does not necessarily directly imply that the stable density stratification increases. Stratification supports destabilizing shear to the point of marginal stability. This shear is largely imposed by internal waves supported by the same stratification, and which thus may help destroy their own habitat. As a result, we cannot yet predict the effects of ocean warming (or cooling) on vertical diapycnal exchange of suspended and resolved materials. Long-lived prejudice on particular ocean mechanisms hamper studies on potential coupling mechanisms, as it is quite likely that mechanisms on the large ocean basin scales (circulation), on meso-scales (eddies) and small-scales (turbulence) are all more or less connected.

Future Challenges

The ocean's governing physical processes of transport and redistribution of suspended and resolved materials are essentially three-dimensional at scales varying from mm to 1000's of km spatially and from 0.01 s to years in time. Technically we still have a long way to go from present-day mainly 1D observation, even though satellites provide 2D imaging from the ocean surface. While the ocean's aspect ratio is approximately 1:1000, given the 3,700 m average depth and typical width of basins, meso-scale processes have 1:100 to 1:10 ratios and small-scale processes have scale ratios >1:10 and into 1:1, the essential ratio for fully developed turbulence. To understand relevant ocean processes, we require information on the precise coupling between all scales. We need this information from observations that go beyond the 1D, at least into 2D, preferably full 3D, and over long time scales while resolving the smallest time-scales.

One development is the installation of cabled networks at the ocean floor, with instruments that are powered from shore where they also send their data to that can be monitored in quasi-real-time. This could be ideal for potential public participation research to cope with the expected large data sets, provided it is well guided. While the two principal means of investigating flows, at a fixed position in space "Eulerian," after the Swiss mathematician L. Euler, or following a particle and keeping track of its position "Lagrangean," after the French mathematician J. L. Lagrange, are defined since the Eighteenth century, practical sampling of physical oceanographic processes provides a more or less blurred image on the dynamics.

Moorings should not move to be Eulerian, which is physically impossible because the main drag and buoyancy forces do not balance as they are near-perpendicular to each other. During careful planning including minimizing drag by using thin cables and maximizing buoyancy one could arrive at weakly moving moorings by the large-scale current only, so that the Eulerian concept is well enough approximated.

Shipborne profiling, by the lowering of an instrument package at typical speeds of 0.5–1 m s^{-1} and mapping into quasi-2D by grouping several CTD-stations together in an ensemble requires a rather severe transformation or assumption of quasi-synopticity to study basically steady-state or limited dynamical processes. Similarly, data sampled via autonomous vehicles cruising the oceans at typical speeds of about 0.25 m s^{-1}, which are in the same range as general ocean flow speeds, can provide a cross-sectional overview of ocean properties but are difficult to transform to either Eulerian of Lagrangean frames of reference. Given their complex technical support needed it is not foreseen that such vehicle sampling will invalidate shipborne research in the near-future, also because ships remain needed to deploy other oceanographic equipment like moored instrumentation.

Although some success has been obtained using satellite-tracked surface drifters to examine flows in a Lagrangean frame, a major challenge is the set-up of an ocean-interior Lagrangean experiment in which (a large set of) particles is followed, e.g. via a long baseline acoustic tracking system. While this is presently limited to small spatial scales only and perhaps in the future extendable to meso-scales, large ocean-basin scales are not expected to be resolved thus. Presently, the ocean-basin scales are resolved by an impressive network of some 3,800 Argo drifters, but these drifters are only tracked (via satellite) when they transmit their data after re-surfacing, every few days. Such observations are important for monitoring the state of the ocean, but are difficult to be used to understand the dynamical influence of the small- and meso-scales on the large-scale, in a proper Lagrangean framework.

While the all-scales research may prove a too large challenge observationally, the next step down (or up) is to study processes to such extent that they may be properly parameterized in ocean modeling. Such parametrizations can only be based on a thorough knowledge of the underlying physical processes. The advancement of computing power is so rapid that, still on relatively small scales, robust DNS modeling can progressively handle larger scales and more complex flows. (Direct Numerical Simulation numerically solves the full NS-equations at all scales albeit up to the computational limits, without a turbulence model). Nevertheless, the large Reynolds numbers typical for most ocean regions and implying the ubiquity of ocean turbulence are still (far?-) future challenge for most "realistic geometry" DNS. Work is also still in progress on the development of sub-grid scale DNS-models into less computationally cost-demanding LES models that resolve only the large (eddy) scales. Perhaps a better set-up is to incorporate first LES into regional seas and large-scale ocean circulation models, as the effects of ocean eddies are not well understood on heat transport, for instance. Yet, simple nu merical and analytical model analyses are needed to understand numerous existing small- and meso-scale problems.

The precise transfer of energy from linear to nonlinear (internal) waves and their breaking is not solved. Boundary dynamics above sloping topography and wave breaking are different from frictional flow models above flat bottoms. The same for interaction between meso-scale eddies and their effect on ocean heat transport. While principal energy sources that drive the ocean are known, like the Sun, Earth rotation and tides, their precise interaction processes are key outstanding problems. The occurrence of extreme events has to be taken further than a statistical description. Such problems are not only challenging for physical oceanography, but also for the impact on marine life and other disciplines.

The ocean seems vast and unknown, but although we do not fully understand its capability of feed-back systems to varying impacts there is a tendency toward general consensus that it is vulnerable. The ocean is not only vulnerable to pollution, in direct manners like plastic dumping and overfishing or in indirect manners like artificial heating, it is likely vulnerable to unknowns like ice sheet collapse and human extraction of (tidal) energy. Better knowledge of governing physical oceanographic processes is needed to argument against persistent prejudices. Some say the water motion contains "a lot of energy." However, there is relatively little energy in tides, by far not enough to cover human power-consumption, but it is crucial for various effects ranging from the maintenance of the ocean stratification and overturning circulation, via diapycnal turbulent mixing, to most important fishery nurseries in estuaries. Some say the Gulf Stream may change direction. However, the main drivers are not the ocean density variations but the Earth rotation and the wind, which will not be significantly saltered on short time scales.

Ocean Chemistry

Chemical oceanography deals with the conditions and interactions between organic and inorganic compounds and the biological, physical and geological conditions in the sea. The interdisciplinary scientific branch also studies the interactions of chemical systems with the marine organisms, the atmosphere, the rocks and the sediments and the changes caused by oceanic mixing processes.

The basis of this observation is based on the distribution and dynamics of chemical elements and isotopes and molecules. This ranges from basic physical, thermodynamic and kinetic chemistry towards the interactions of marine chemistry with biological, geological and physical processes. This includes both inorganic and organic chemistry as well as studies of atmospheric and terrestrial processes. In addition, the Chemical Oceanography processes that occur on a variety of spatial and temporal scales.

The marine chemistry deals with the chemical composition of sea water; it is therefore a special field of water chemistry. As part of the hydrosphere, the chemical oceanography also deals with aspects of atmospheric chemistry, geochemistry, the biosphere and environmental chemistry.

Major Ions of Seawater

Major ions are defined as those elements whose seawater concentration is greater than 1 ppm. The main reason this definition is used is because salinity is reported to ± 0.001 or 1 ppm. Thus, the major ions are those ions that contribute significantly to the salinity. According to this definition there are 11 major ions. At a salinity of S = 35.000 seawater has the composition given in Table below. The data in this table mainly come from a set of complete major ion analyses conducted at the University of Liverpool in the early 1960s on a representative set of over 100 seawater samples. Such a major effort will probably never be undertaken again baring some major breakthroughs in analytical chemistry. Using the concentrations units of g kg-1 allows us to determine the contribution to salinity. Chemists prefer to use moles for units, thus mmol kg^{-1} are preferred (e.g. rather than mmol l^{-1}). Also mmol kg^{-1} is preferred over mmol l^{-1} (or mM) because one kg of seawater is the same at all values of T and P, wheras the mass in one liter can vary.

Table: Concentration of the Major Ions.

	At salinity (PSS 1978):S = 35.000%			
	mg kg^{-1} S^{-1}	g/kg	mmol/kg	mM
Na^+	308.0	10.781	468.96	480.57
K^+	11.40	0.399	10.21	10.46
Mg^{++}	36.69	1.284	52.83	54.14
Ca^{++}	11.77	0.4119	10.28	10.53
Sr^{++}	0.227	0.00794	0.0906	0.0928
Cl^-	552.94	19.353	545.88	559.40
SO_4^-	77.49	2.712	28.23	28.93
HCO_3^-	3.60	0.126	2.06	2.11
Br^-	1.923	0.0673	0.844	0.865
$B(OH)_3$	0.735	0.0257	0.416	0.423
F^-	0.037	0.00130	0.68	0.070
Totals	1004.81	35.169	1119.87	1147.59
Alkalinity	-.-	-.-	2.32	2.38
Everything else	-.-	~0.03	-.-	-.-
Water	-.-	~964.80	~53,555.	~54,881

Seawater is first of all a solution of NaCl. Na and Cl account for greater than 86% of the salt content by mass. The order of the other cations is Mg^{2+} > Ca^{2+} > K^+ > Sr^{2+}. The anion Cl- is approximately equal to the sum of the cations. The other anions (SO_4^{2-}, HCO_3^-, Br-, F-) are much less significant in the charge balance of seawater.

An element is conservative in seawater if its ratio to salinity is constant in different parts of the ocean. The ratio of one conservative element to another will also be constant. One way to establish if an element of unknown reactivity is conservative is to plot it versus another conservative element or conservative property like potential temperature or salinity. This is referred to as The Law of Constant Proportions. Conservative elements are conservative because they have very low chemical reactivity and their distributions in the ocean interior are determined only by currents and mixing. The list of elements considered conservative has changed over time as analytical techniques have improved.

The Law of Constant proportions breaks down some places where water sources have different ionic ratios or where extensive chemical reactions modify the composition. Examples are:

- Estuaries: The average composition of river water is given in Table. The concentrations are given in mg l^{-1} and can be compared with seawater concentrations. The main difference is that HCO_3^- is the main anion in river water and has a much higher concentration than Cl^- (which is the lowest of the major anions in river water). Calcium is the main cation in river water, followed by Na and Mg, then K.

- Evaporitic brines form in isolated seawater embayments where various solids precipitate, thus changing the relative concentrations left in solution.

- Anoxic waters in restricted marine basins and pore (interstitial) waters in sediments where dissolution/precipitation and oxidation/reduction reactions can change the relative composition (especially of SO_4^{2-}).

- Hydrothermal vents: These high-temperature waters differ from seawater in that Mg, SO_4 and alkalinity have all been quantitatively removed.

Specific Examples

(a) Na, K, SO_4, Br, B and F have constant ratios to Cl and each other, everywhere in the ocean. These elements are conservative. There are reasons to think that SO_4 may be non-conservative in anoxic marine basins like the Black Sea but conclusive results showing non-conservative distributions have not been observed. Sulfate depletions are frequently seen in pore water due to sulfate reduction. Boron is partly present in seawater as the neutral species ($B(OH)_3^o$) and it has been hypothesized that B may be distilled from surface water in the tropics during evaporation and transported to high latitudes through the atmosphere. Rain is enriched in B but non-conservative distributions in the ocean have not been identified.

(b) Ca has small (+ 0.5%) but systematic variations within the ocean. This has been known since the earliest analyses of seawater. Dittmar used precise analyses of 77 Challenger samples to show that the calcium/chloride ratio was higher in deep water then in surface waters.

When the Ca increase was first discovered by Dittmar it was hypothesized to be due to dissolution of $CaCO_3$ particles. Brewer et al showed that the change in alkalinity with depth in the ocean (where Alkalinity $\approx HCO_3^- + 2CO_3^{2-}$) was less than that expected for the change in Ca. Actually according to the $CaCO_3$ solubility reaction (e.g. $CaCO_3(s) = Ca^{2+} + CO_3^{2-}$), the changes should follow the expression ΔAlkalinity = 2ΔCa. Calcium increases by 100-130 µM as deep water flows from the Atlantic to the Pacific but alkalinity only increases by 120-130 μM. As the calcium data are probably sound, Brewer et al. suggested that this was because the alkalinity was low due to titration by HNO_3 produced by respiration of organic matter in the deep sea. The correct comparison should be of Ca with potential alkalinity which the total alkalinity is corrected for the NO_3 produced according to:

$$\Delta\text{Potential Alkalinity} = \Delta\text{Alkalinity} + \Delta NO_3^-$$

The increase in NO_3 from the deep Atlantic to Pacific is about 30 μM, which should decrease alkalinity by the same amount. This is an important correction to make but there is still a "calcium problem".

New data from de Villiers (personal communication) shows that Ca increases systematically from the surface to the deep water and from the North Atlantic to North Pacific. $CaCO_3$ dissolution may not be the only source of dissolved Ca. DeVilliers showed that variations in Ca and potential alkalinity were in good agreement from 0-1000 m and in the deep water (>3500 m), but that there was additional excess Ca in the mid-water column centered at about 2000 to 2500 m. She argued that this was primarily due to diffuse source low-temperature hydrothermal input from mid-ocean ridges. For support the xsCa correlates well with other hydrothermal tracers like [3]He and Si, and low Mg.

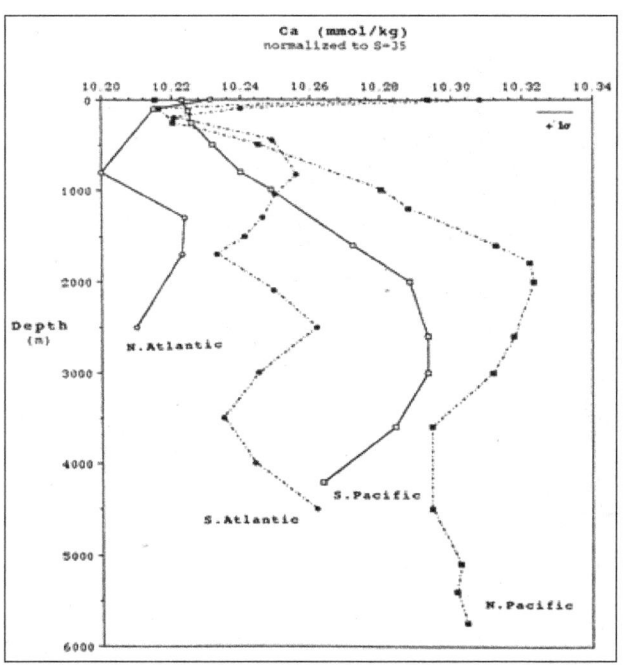

Figure: Ocean Ca Profiles.

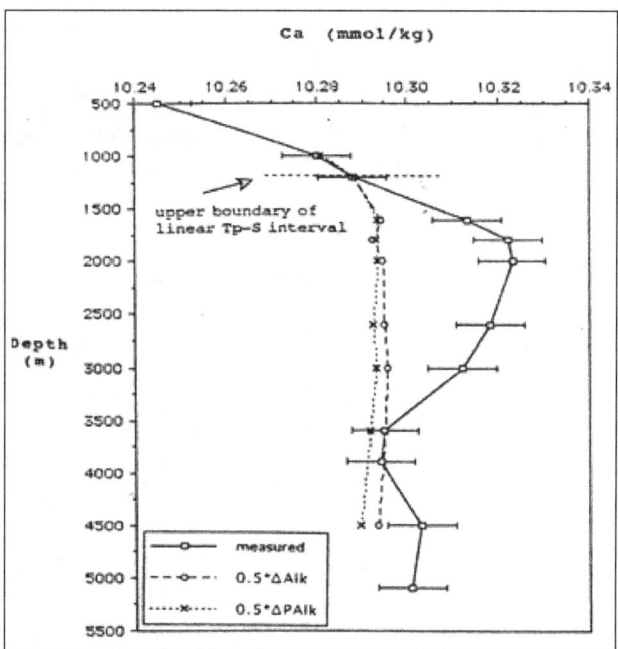

Figure: Ca and Alkalinity.

(c) Sr also increases from the surface to the deep water and from the North Atlantic to the North Pacific. The deep water is about 2% enriched relative to the surface water. The nutrient element PO_4 has a similar pattern and there is an excellent correlation of Sr with PO_4 in both surface waters and with depth confirming that Sr shows a nutrient like pattern. The biogenic mineral phase Celestite ($SrSO_4$) has been proposed to be the transported phase. A link has been established with the microzooplankton, Acantharia. Acantharians are marine planktonic protozoans and they are the only marine organisms that use Sr as a major structural component. They make their shells and cysts out of Celestite (SrSO$_4$) even though seawater is extremely undersaturated. Acantharians and their cysts are present in most of the world's oceans and are frequently more abundant than their protozoan counterparts, the radiolarians and the foraminifera. Acantharian cycts have been recovered from sediment traps and their concentrations decrease sharply with depth over the top 1000m mirroring the increase in dissolved Sr in the water.

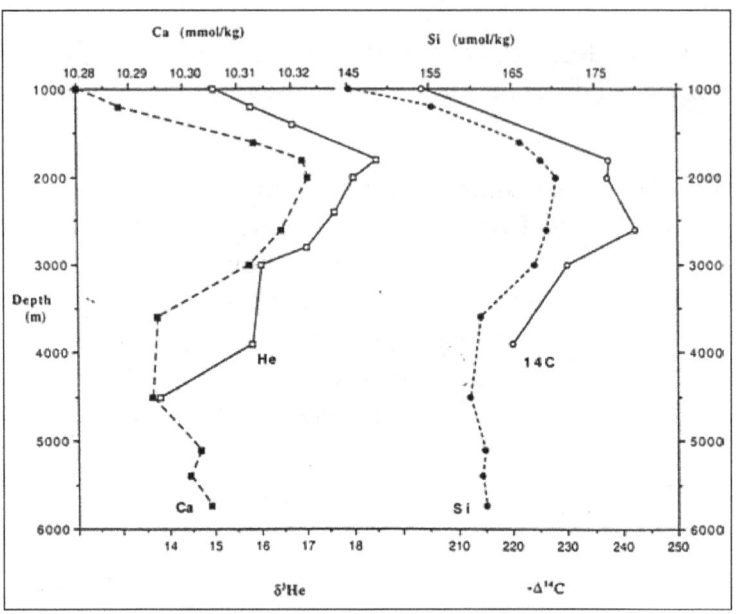

Figure: Comparison of Ca, He, Si and ^{14}C.

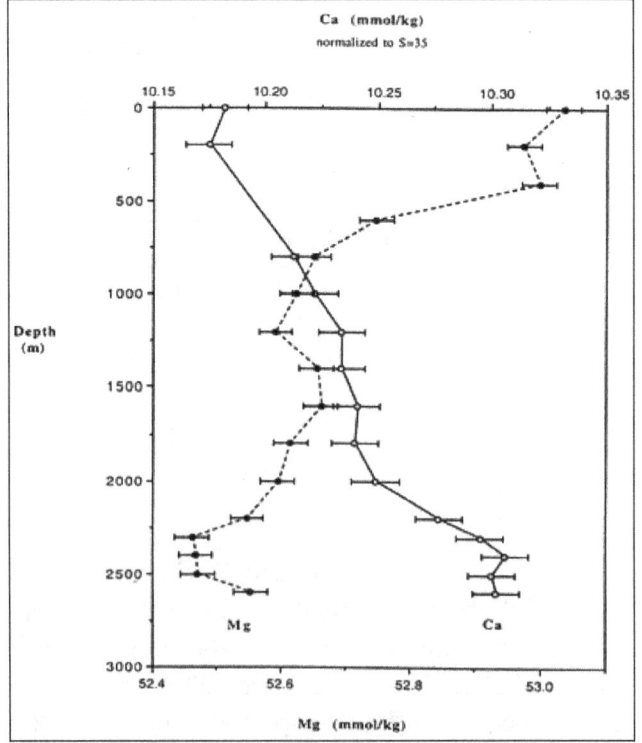

Figure: Comparison of Ca and Mg. Both normalized to S = 35.0

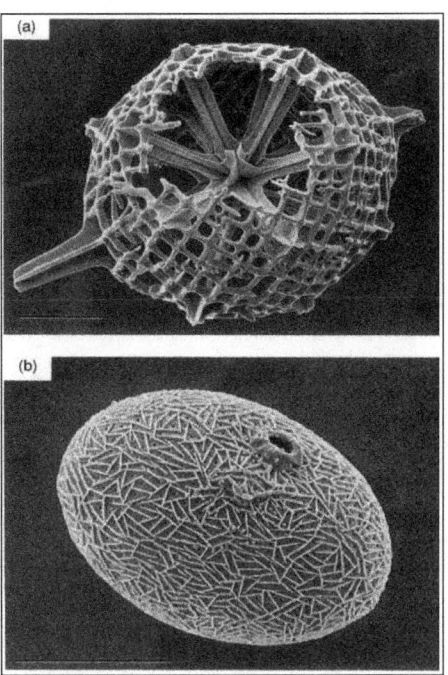

Figure: (a) An Acantharian skeleton. The skeleton is broken revealing a number of the twenty equidistant spines characteristic of this protozoan group. Reference bar= 50 μm. (b) An Acantharian cyst. This cyst is oriented to show the large pore present in many cysts. Reference bar= 50 μm.

The long residence time of Sr in the ocean could suggest that its concentration does not vary over Quaternary time scales, but in fact the best estimates of the modern Sr budget suggest that it is far from steady state. And in fact large changes (up to 12%) in Sr/Ca in planktonic foraminifera have been observed over the past 150 ka (glacial/interglacial time scales) suggesting that Sr/Ca was higher during glacial maxima. This variability reflects changes in sea level, river fluxes and carbonate accumulation rates.

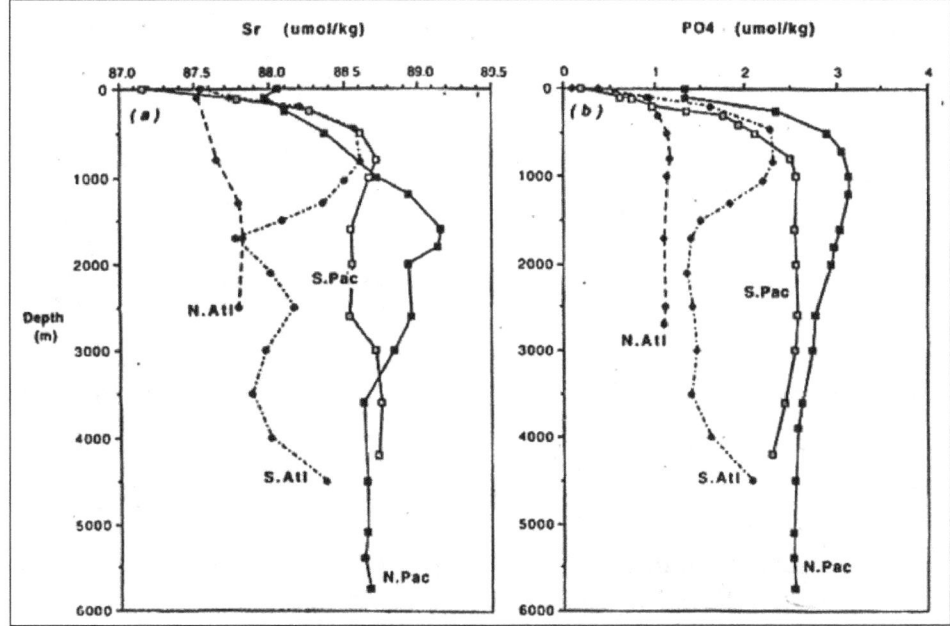

Figure: Sr and PO$_4$ profiles from the Oceans.

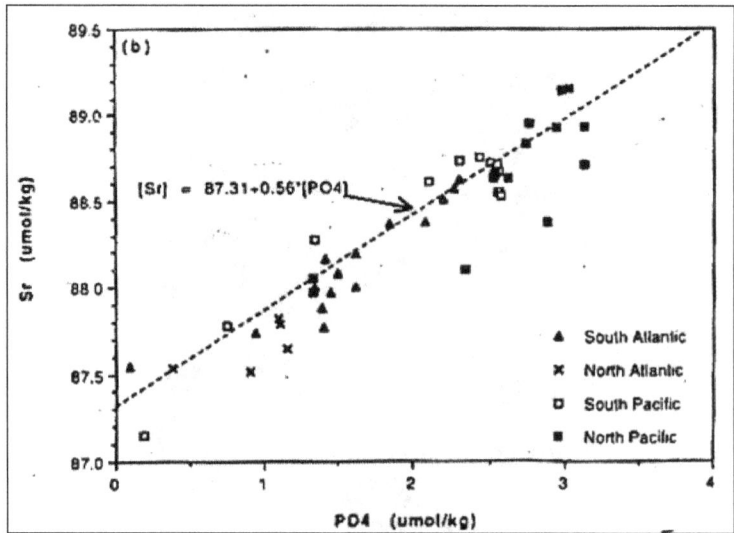

Figure: Sr – PO$_4$ correlations for the Pacific Ocean.

Sr has also been attractive as a proxy in paleoceanographic studies because of its long residence time in the ocean (5Ma), which implies a uniform distribution and conservative nature. Sr is taken up by corals and the coral Sr/Ca ratios have been used to infer variability in sea surface temperature. An example SST - Sr/Ca correlation is shown in figure. Beck et al and Guilderson et al used Sr/Ca results from corals to suggest that the tropical western Pacific and western Atlantic were 4°C and 6°C cooler during the last glacial maximum (LGM) than today.

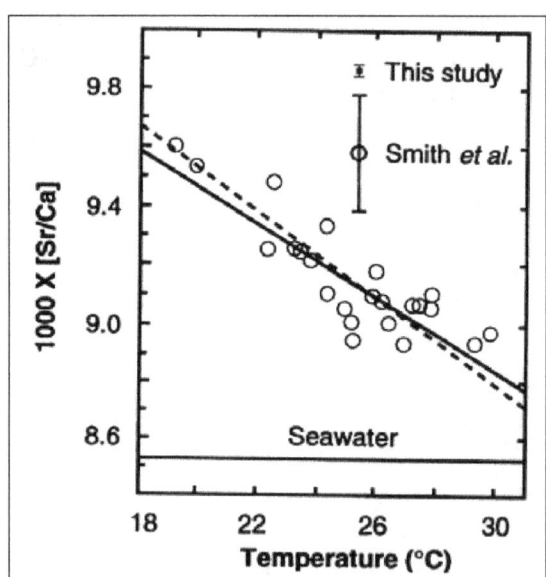

Figure: Plot OF Sr/Ca ratios in scleractinian corals against the temperature of the seawater from which they grew.

This approach has to be used with care as interspecies differences and effects of growth rate can also affect the Sr/Ca ratio. Sr turns out to be difficult to use as a proxy for SST as its partitioning has multiple controls. Stoll et al analyzed the Sr/Ca ratio in planktonic foraminifera for the past 150 ka and found variations of up to 12% on glacial / interglacial timescales. At least some of this variability was interpreted in terms of sea level changes, together with large changes in river input and carbonate sediment accumulation.

(d) Until recently Mg was thought to be conservative. Its residence time in seawater (13 Ma) is much longer than that of Sr (5Ma) or Ca (1 Ma). Again, de Villiers (personal communication) has recently found relatively large Mg anomalies in deep waters located over mid-ocean ridges. An example of vertical profiles of Ca and Mg above the East Pacific Rise at 17°S 113°W show that depletions in Mg mirror increases in Ca. Mg is known to be totally removed in high temperature hydrothermal vent solutions. However, diffuse low-temperature hydrothermal solutions are thought to be 10x to 100x more important for chemical fluxes. Unfortunately these end member concentrations have not been well defined.

Mg is also taken up by foraminifer's shells. The tropical planktonic foraminifera Globigerinoides sacculifer is a popular sample for such studies. Experimental studies by Lea et al demonstrate the potential of Mg/Ca as a paleothermometer. The response of Mg/Ca to temperature is stronger than that for Sr/Ca. Lea et al used the historical Mg/Ca record from equatorial sediments to postulate that sea surface temperature was lower by about 3°C in that region during the last glacial period and that the increase in tropical SST led Greenland warming during the Bolling Transition at the end of the last glacial period (about 14.6 thousand years ago). The time lags in such records remain controversial as other paleo-SST records suggest close synchronous SST change between tropical ocean regions in the Pacific and Greenland.

The deep-sea temperature record for the past 50 million years has been produced from the Mg/Ca ratio in benthic foraminifera calcite. This record suggests a cooling of ~12°C over the past 50 My in the deep-sea. When combined with the simultaneous measurement of benthic $\delta^{18}O$, the Mg record provides estimates of global ice volume. The data suggest that the first major continental-scale ice accumulation occurred in the earliest Oligicene (34 Ma). DIC (H_2CO_3 + HCO_3 + CO_3) varies by < 20% with depth in the ocean due to vertical transport and remineralization of both $CaCO_3$ and organic matter.

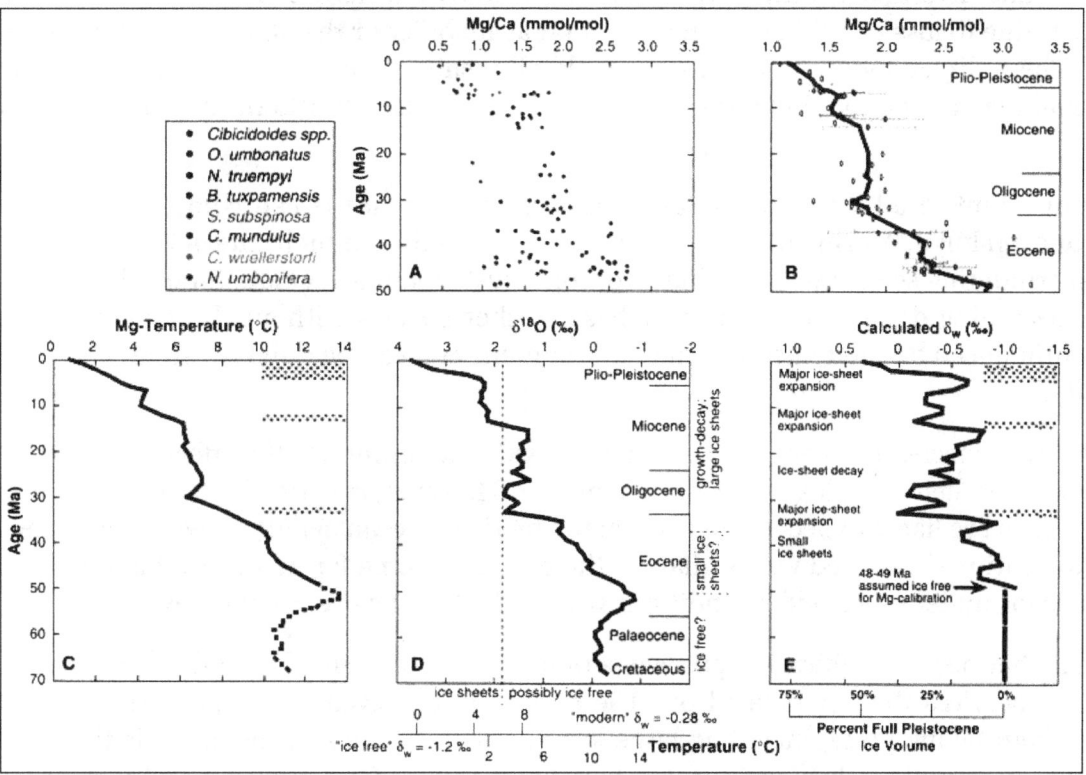

Figure: Mg/Ca as a Temperature tracer. When used with $\delta^{18}O$ can be used to estimate variations in ice volume.

Acidity is measured on a pH scale, where lower pH indicates more acidic water. Ocean pH has dropped by thirty percent globally during the last two hundred years. Even though the drop in pH appears small (from 8.2 to 8.1), the pH scale is logarithmic, meaning that this change is large enough that it may already be beginning to affect some of the oceans most beloved and biologically important residents, including corals.

The changing acidity of the oceans threatens to throw off the delicate chemical balance upon which marine life depends for survival. The scant attention this issue has received has focused primarily on corals, which are threatened with extinction within this century unless we change course. Corals are the framework builders of reefs, by far the most diverse ecosystems of our oceans. However, the effects of acidification are not going to stop with reefs, like dominoes, the impacts are going to be far-reaching throughout the oceans.

The increased acidity in the oceans is expected to lead to a shortage of carbonate, a key building block that some animals (and plants) need to build their shells and skeletons. Besides corals, other animals that use carbonate to build and strengthen their shells (called calcifiers) may also suffer. These animals include shellfish like clams, oysters, crabs, and lobsters.

But the problem is not just about carbonate. Marine life ranging from the smallest plankton to the largest whale may be affected by ocean acidification, with direct and indirect impacts. Direct effects on the physiology of animals and plants may result from a lack of available carbonate needed to build and strengthen shells and skeletons. Within this century, increasing acidity may reduce the ability of certain fish to breathe, and increase the growth rates of some sea stars. In addition, it is likely to inhibit some fish from smelling cues that direct them to suitable habitat or away from predators, and divert energy away from important survival tasks of many species. Indirect effects may occur due to loss of habitat, changes in food availability or the abundance of predator populations. These direct and indirect impacts could all result in animals becoming less fit to survive which could lead to population level consequences and ripple effects throughout marine ecosystems.

pH is important to all living things and changes in internal pH can compromise an animal's health or even kill it. Marine plants and animals must maintain their internal pH relative to that of the surrounding seawater. Some have complex systems that regulate internal pH, preventing it from becoming dangerously acidic or basic. Other species without these systems are more heavily influenced by their surrounding environment and can be quickly threatened by changes in acidity.

Some marine species, such as fish, were once considered immune to the effects of ocean acidification because of their complex buffering capabilities. However, research is showing that even some marine fishes are likely to become overwhelmed by rising ocean acidity. If we allow ocean acidification to continue unabated we risk having the oceans become far less vibrant and dynamic. The future oceans under more acidic conditions will not be like the oceans of today.

It is clear that ocean acidification poses a serious threat to marine life and to the ocean-derived goods and services that we depend on. The only way to prevent these potentially catastrophic changes from taking place, including large losses of coral reefs and the animals that depend on them, is to drastically reduce our carbon dioxide emissions. Our actions over the next few years will determine how acidic the oceans will become. Without any changes the oceans are expected to become more acidic in the coming decades than at any time in at least the last 20 million years. The speed at which we are changing ocean acidity is unprecedented. This is highly concerning because most marine organisms living today have never adjusted to such rapid changes in pH 6 Luckily, this does not have to be the oceans' fate. We must reduce the risk of catastrophe by quickly and comprehensively reducing our carbon dioxide emissions.

Impacts

Pacific Oyster Farms Fail

Shellfish farmers may already be feeling the devastating impacts of ocean acidification on their livelihood, as they have been experiencing difficulties raising oyster larvae since 2005. This is likely due at least in part to more acidic conditions along the U.S. Pacific Coast. These losses are reportedly having drastic consequences for the 111 million dollar oyster industry of the West Coast.

Other mollusks, including mussels, scallops and clams, also appear to be extremely vulnerable to increasing acidity. These animals all create calcium carbonate shells to protect their soft bodies from predators, disease and harsh ocean conditions. Mollusks create their shells out of a highly soluble form of calcium carbonate. Severe declines in shell growth rate are likely by the end of this century.

Slower shell growth is likely to reduce the ability of mollusks to survive, which would have significant impacts on commercial fisheries. One study suggests that if this slowed growth predicted for 2100 had occurred in 2006, mollusk fisheries would have lost between 75-187 million dollars. This prediction is likely to be an underestimate since it does not take into account other negative impacts of rising acidity, such as reduced reproductive success and larval survival.

Corals Show Decreased Growth

Coral reefs are very important to many coastal communities and national economies, and they too have started to show signs of decline that may be due to ocean acidification. Some of the largest reef-building corals on the Great Barrier Reef are showing more than fourteen percent reductions in skeletal growth since 1990, the largest decrease in growth rate in the last 400 years.11 Declines in growth rate could result in a mass die-off of tropical coral reefs by the middle to end of this century.

Some 500 million people worldwide depend on reefs for coastal protection, food and income. Economists value reefs at between 30 to 172 billion dollars per year. Healthy reefs provide goods and services to society, including fisheries, coastal protection, tourism, education and aesthetic values. In Hawaii alone, coral reefs annually generate some 364 million dollars through tourism. If reefs collapse because of rising acidity, global warming and other threats, coastal communities will bear the brunt of these losses. Serious health consequences could ensue for the estimated 30 million people who rely almost solely on coral reef ecosystems for protein and protection. The potential losses from a decline in coral reefs will be felt from the smallest coastal subsistence communities all the way up through the global economy.

Coral Reef Habitat Loss

Coral Reefs

Coral reefs are probably home to at least a quarter of the entire biological diversity of the oceans, a seemingly limitless number of species, and serve as some of the most beautiful habitats in the world. Although they cover just over one percent of the world's continental shelves, coral reefs serve as important habitat to as many as one to three million species, including more than twenty-five percent of all marine fish species. These millions of species feed, reproduce, shelter larvae and take refuge from predators in the vast three-dimensional framework offered by coral reefs.

By the middle of this century, if carbon dioxide emissions continue unabated, coral reefs could be eroding through natural processes faster than they can grow their skeletons due to the combined pressures of increasing acidity and global warming. Reefs may become nothing more than eroded rock platforms, greatly changed from the structures that so many species rely upon for habitat. Corals face severe declines and even extinction, which will in turn threaten the survival of reef dependent species.

Fish

As many as four thousand species of fish depend on coral reefs for habitat. Some of these fish, such as butterfly fish, feed exclusively on the coral itself. Other species of fish also depend on coral reefs as sources of food, shelter and nurseries. Loss of coral reef dependent fish can be expected as reef habitats become less available.

Extensive die-offs of coral due to bleaching events can serve as an example of how interrelated coral reef fish are with their coral habitats. After one event in Papua New Guinea, 75 percent of the coral reef fish species declined in abundance and several species even went extinct. Unfortunately, these examples give us a clue about what we might expect in the future as reefs begin to disappear due to rising ocean acidity.

Sea Turtles

Sea turtles are some of the most endangered marine animals and are often found resting and feeding within coral reefs. They like to feed on reef species such as sponges, mollusks, algae and soft corals. Declines in coral reefs could impact turtle feeding behaviors and could cause them to turn to less nutritious food sources or even go hungry.

The ability for a turtle to dig a nest and successfully incubate eggs has also been connected to healthy coral reefs, as these activities are in part connected to the type and amount of sand on nesting beaches. Beach sand near coral reefs is most often made up of the skeletal remains of plants and animals that live on the reef, including parts of the reefs themselves. As ocean acidification worsens the abundance of reef species will likely diminish, which could result in changes to the type and amount of sand reaching nearby beaches. Changes in the makeup of the sands on nesting beaches could negatively impact sea turtles' ability to successfully produce new hatchlings and could reduce the population size of these already endangered species.

Food Web Disruption

Pteropods

Pteropods, abundant, tiny swimming sea snails are sometimes referred to as the "potato chips of the sea" because of their importance as a food source for so many species. They can reach densities of thousands of individuals per cubic meter, and are particularly important in the polar and sub-polar food webs and serve as an important part of the diets of zooplankton, salmon, herring, birds and baleen whales. Pteropods build calcium carbonate shells, a process that is particularly vulnerable to increasing ocean acidity. As early as the year 2050, pteropods may be unable to form these shells, threatening their ability to survive. If they cannot adapt to living in more acidic waters, their populations will plummet, which could result in ripple effects throughout the food webs that depend on them.

Salmon

The high-latitude seas are home to some of the most lucrative fisheries on the planet, fisheries that are heavily dependent upon pteropods and other marine calcifiers for food. Some areas of the Arctic Ocean are already experiencing corrosive conditions that within the decade could be more common throughout the high-latitude waters of both the North and South. By the end of this century surface waters of the Arctic and Southern Oceans and parts of the Northern Pacific will be corrosive to marine calcifiers, with obvious cascading effects for the fisheries that rely on them.

North Pacific salmon depend heavily upon pteropods for food. In fact, pteropods can make up 45 percent of the diet of juvenile pink salmon. The North Pacific salmon fisheries provided three billion dollars worth of personal income to fishermen and others in 2007, and supported 35 thousand jobs in just the harvesting and processing of the fish. Other commercially important fish species that eat pteropods include mackerel, herring and cod—all of which could risk collapse if pteropod populations decline.

Killer Whales

One of the most devastating impacts of rising ocean acidity could be the collapse of food webs. Declines in the smallest of species, like pteropods and other plankton, could reverberate throughout the oceans, ultimately impacting the largest marine species. The Chukchi and Northern Bering Seas are some of the richest fishing grounds in the oceans and are home to predators as varied as gray whales, seals, sea ducks and walruses that all depend on marine calcifiers for food. By the end of this century parts of these seas will be inhospitable to many shell-making organisms, which could result in food web collapse.

For example, resident killer whales in the North Pacific prefer to eat salmon, in fact 96% of some killer whales diet is made up of salmon. When the base of the food web disappears, the effects can travel all the way to the top. If predators are unable to supplement their diets with other food sources, food webs may even collapse entirely. Top predators like the emblematic killer whale could suffer, which in turn could have further implications. This iconic species and important tourist attraction could be threatened by cascading impacts from the loss of pteropods.

Acidification may Affect Shellfish

Sea Urchins

Sea urchins are found in various environments, from coral reefs to rocky coasts, and are crucial grazers in any environment. Sea urchins on a reef help to protect the reef by eating some of the algae that might otherwise encroach on corals and displace them from the reef. Sea urchins are also the target of a lucrative fishery that brought in nearly six million dollars in revenue for the state of Maine in 2009.

Sea urchins reproduce by releasing eggs and sperm directly into the surrounding seawater. Under acidified conditions the sperm of some sea urchins swim more slowly, which reduces their chances of finding and fertilizing an egg, forming an embryo and developing into sea urchin larvae. Even under normal conditions, only a small percentage of sperm are able to locate and fertilize the eggs. The majority of sea urchin embryos and larvae are eaten by fish and only a few survivors mature into adults. To make up for this low success rate, sea urchins normally release millions of eggs and sperm into the surrounding water. However, the more acidic conditions predicted for the end of this century could reduce the number of sperm certain species release, thereby further decreasing the size of the next generation of sea urchins.

Sea urchins are in the category referred to as calcifiers since they create their skeletons, called tests, out of calcium carbonate. Like many other calcifiers, such as corals, pteropods, and oysters, sea urchins are likely to find it more difficult to build their calcium carbonate skeletons in an acidified ocean. Young sea urchins have been observed to grow slower and have thinner, smaller, misshapen protective shells when raised in acidified conditions, like those expected to exist by the year 2100. Slower growth rates and deformed shells may leave urchins more vulnerable to predators and decrease their ability to survive.

Lobsters

Lobsters are long-lived and are found through all the oceans of the world. Under more acidic conditions, their larvae had lighter, less dense shells, which could make them more susceptible to predation and less able to survive. Declines in lobster populations may also result from losses in food sources, such as sea urchins, which are also vulnerable to increased acidity.

Lobsters and other crustaceans such as crabs periodically molt throughout their lives and grow new shells out of calcium carbonate and chitin. They use the calcium carbonate to harden their shells in a very different process than the one used by other marine calcifiers. This process of calcification is very important as hard exoskeletons protect lobsters, crabs and other crustaceans from predators and disease. The creation of a new shell requires a lot of energy—energy that is also needed for other functions like growing, breathing and reproducing. Increasing ocean acidity may drive some lobsters to create larger shells. The exact reason for this is unknown; however, it is likely to have negative consequences as it will almost certainly result in energy being diverted away from other activities that are vital to survival.

Decreased survival of lobsters and crabs due to increased predation and abnormal shell growth could have serious implications for the economies of many local communities. In 2008, the U.S. lobster fisheries alone brought in an estimated 300 million dollars. Crabs also provide an important part of the fishery for many coastal states, including Maryland, where half of the total fisheries catch is often made up of blue crabs.

Brittle Stars

Brittle stars play a crucial role in their environment as burrowers and as a food source for larger predators, like flatfish. Burrowing stirs up sediment and allows oxygen to mix with it, which is very important to many bottom dwelling species. Brittle stars, like sea stars, have a calcium carbonate skeleton made up of many small plates held together by skin and tissue. As their name implies, brittle stars are quite fragile, but this trait is usually to their advantage. The spindly arms of a brittle star break off when the animal senses danger and, under normal conditions, can quickly regenerate. However, this process is likely to be disrupted in the future as the oceans become more acidic.

Studies have shown that brittle stars can still regenerate their arms under acidic conditions but do so with less muscle mass than usual. It appears the brittle stars are sacrificing building muscle in order to create the calcium carbonate parts of their arms. The brittle stars not only created insufficient amounts of muscle for their new arms to function correctly, they also devoured muscle from their already existing arms to provide energy for the now much harder process of building calcium carbonate. Muscle is essential for movement, finding food, and burrowing in sediment, which are all tasks vital to survival. Weakened arms could decrease the ability of brittle stars to survive in a more acidic ocean.

Increased acidity is also likely to threaten brittle star larvae. By the middle of this century it is likely that they will have difficulty developing and many, if not all, are expected to die after exposure to acidified conditions. Thus, brittle stars appear to be very vulnerable to increasing ocean acidity both as adults and larvae, which could result in severe population declines in the future.

Sea Stars

Sea stars help to maintain diversity on a reef by acting as important predators, keeping the populations of other species in check. Sea stars do not have a continuous skeleton made up of calcium carbonate, rather they have hundreds of tiny calcium carbonate plates embedded within their tissue. These plates tend to make up a relatively small percentage of a sea star's total body weight. This may explain why sea stars appear to respond differently to increased ocean acidity than other marine calcifiers that have larger, more continuous calcium carbonate shells and skeletons, like corals and oysters. Studies on sea stars have shown varied responses to increased acidity; for example, one study found that the purple sea star decreases calcification but increases its overall growth. This increase in growth rate could lead to increased feeding rates, putting more pressure on preferred food sources (like mussels) and therefore causing population declines among mussels. As mussel populations are likely to have already declined because of ocean acidification, sea stars could be forced to switch to other food sources or suffer population declines themselves if they cannot find suitable substitutes. Increasing growth rates among sea stars could cause large changes through the ecosystems that sea stars have traditionally kept in balance.

Another type of sea star, the sun star, has also been found to increase its growth rate in acidic conditions and was in fact found to perform better due to positive effects to its metabolism, which could decrease the time it takes for its larvae to develop. The sun star is an important predator in its food web and influences how abundant many species are. Decreased larval development times could mean that larvae develop into juveniles earlier, but this new timing could be out of sync with the factors that the survival of juveniles are reliant upon, such as water temperature, food availability and lack of predators. Developing early could put juveniles in adverse conditions that prevent them from growing properly or surviving. Ecosystems' exquisite balance is determined by the interconnectedness of many species relying upon each other's development and abundance. Small changes can put these systems out of balance, and while changes may appear advantageous in a lab setting they may be harmful in the natural environment.

Acidification may Affect Animals without Shells or Skeletons

Squid

Squid are the fastest marine invertebrates, propelling themselves in speedy bursts at up to 25 miles an hour. This form of high-energy jet propulsion consumes large amounts of oxygen which is sent to the tissues through the blood. Increased ocean acidity is likely to inhibit a squid's ability to transport such large amounts of oxygen, which could impede these fast bursts and inhibit important activities like hunting and avoiding predators. This would inevitably affect a squid's ability to survive. While one study has suggested that cuttlefish, which are similar to squid, show resilience to changes in metabolism from high levels of carbon dioxide, another study identified significant drops in metabolic rates and activity levels for jumbo squid. Clearly more research is needed to better understand the impacts of acidification on these species.

If there were a loss of squid, or population reductions, this could be bad news for many of the commercial species that feed on it, such as king and coho salmon, lingcod and rockfish. Declines in squid populations would also have drastic consequences for squid fisheries, which are very important to California where they are caught and exported all over the world. The fishery for market squid was the most profitable in California in 2008, providing 25 million dollars in revenue.

Clownfish

Larvae of fish that live on coral reefs hatch on the reef and then leave their parents and migrate to the open ocean where they spend the next two to three weeks adrift. When the larvae are ready to return to their home reefs they use their sense of smell and sound to guide them back. Under present day conditions, clownfish can cleverly orient themselves to the smell of home using the scent of sea anemones or the vegetation characteristics of a reef. Unfortunately, under the more acidic conditions predicted for the end of this century, larvae may not be able to discern between the smells of a suitable home and a hostile environment, which could result in their death.

In addition, clownfish larvae also use their sense of smell to avoid predators. In carbon dioxide conditions expected around the end of this century, this smell related predator defense system is disrupted and most returning clownfish larvae are no longer able to discern between predator and non-predator cues. This disruption in the sense of smell may also result in riskier behaviors in clown fish and damselfish. Increased levels of carbon dioxide have been associated with these fish being more active, swimming further away from shelter and not responding to threats such as predators. In studies, five to nine times more fish died because of their risky behavior than those not in acidified conditions. As the oceans become more acidic, larval clownfish could not only be lost at sea, but the conditions may even cause them to swim right into the jaws of their predators.

Damselfish

In addition to smell, some reef fish like the damselfish rely on hearing to find their way back to their home reef. They listen to the noises of a reef using otoliths, which are calcium carbonate structures similar to human ear bones. Using their otoliths, fish larvae can separate the low frequency sounds of breaking waves, currents, and surface winds of the open ocean from the high frequency sounds of gurgling, cracking, and snapping of a coral reef. Damselfish larvae use these distinct noises to navigate back to their home reef and away from the open ocean. Carbon dioxide concentrations expected around the end of this century have been observed to alter the normal development of otoliths in the larvae of an open ocean fish, White Sea bass. If ocean acidification also causes malformation of otoliths in the larvae of reef fishes, such as zebrafish or damselfish, the enhanced otolith growth could make it difficult for the fish to locate appropriate reef habitats and result in population declines. Larger than normal otoliths in damselfish have been shown to decrease their ability to recognize sounds and return to a coral reef. These larger otoliths may cause damselfish larvae to take longer to find the reef and even result in fewer larvae successfully returning to the reef at all. Similarly, in zebrafish abnormal otolith growth has been shown to create serious disruptions in locomotion and balance.

Cardinalfish

Increased levels of carbon dioxide in sea water may decrease the ability of some fish to breathe. Cardinalfish have been found to be particularly vulnerable to increasingly acidic conditions. The ability to take up oxygen decreased by as much as 47 percent in one species of cardinalfish when exposed to carbon dioxide levels similar to those expected by the end of this century. Increasing amounts of carbon dioxide can cause acidification of the blood and tissue, which can lead to death. The cardinalfish may have to divert energy towards mechanisms that balance their internal acidity, leaving less energy for other activities necessary for survival, such as breathing.[95] The reduced ability to breathe will likely have wide-ranging impacts on these fish, including decreased ability to feed, grow and reproduce, which could result in adverse consequences for the sustainability of cardinalfish populations.

As the acidity of the ocean increases, they are simultaneously getting warmer due to climate change. These factors, when combined, may create even more problems than either would create on its own. For example, increased temperature combined with the acidity levels expected by the end of this century proved lethal for one species of cardinalfish tested in the laboratory. These results are particularly concerning, especially if they are found to apply to other marine species, since they show that while individuals might be able to survive one threat, they are less able to endure the simultaneous threats of increasing temperatures and ocean acidification.

Ecological Winners – Algae, Jellies and Invasives may Dominate

There will likely be some species that are able to flourish in an acidified ocean, either because increased carbon dioxide levels benefit them directly or because their competitors are directly harmed by it. The only problem is that the species that appear to be best suited to prosper in high-carbon dioxide conditions are those that we currently see as nuisance, or weedy species. These species are unable to provide adequate habitat and food for the species that we love, and as a result the oceans will look very different if acidification worsens.

Jellies

Jellyfish may be one of the "winners" in a more acidic ocean. It's unclear whether increasing acidity is directly related to the recent increases in jellyfish prevalence, but it does appear that ocean acidification does not harm jellyfish reproduction or the formation of their internal structures, as it does for many other forms of marine life. It is likely that even if acidification is not directly responsible for their recent increased prevalence, it may be creating ocean conditions that are ripe for jellyfish to flourish.

Jellyfish blooms could have disastrous impacts, especially if past disruptions from these creatures are replicated. Previous outbreaks have been responsible for decreasing commercial fish stocks due to competition and predation, as well as harboring various fish parasites. In addition, they also represent a threat to beachgoers and can harm economies that depend on coastal tourism.

Sea Plants

Algae and sea grasses are likely do well in an acidified ocean. These species take up carbon dioxide and sometimes directly compete with calcifiers. So as acidity increases, conditions will likely shift in their favor and they may be able to move into areas where they have not previously flourished.

Observations at natural carbon dioxide vents can give us a glimpse into what the future oceans may look like. These visions show marine communities dominated by green algae and sea grasses. One study of naturally occurring carbon dioxide vents off the coast of Italy found 30 percent lower species diversity, especially of calcifiers, close to the vents compared to the surrounding areas. Some of the calcifiers near the vents had weakened and dissolving shells. The areas near the vents also had higher levels of invasive algal species, including some that are already recognized as stubborn invasives in other marine systems. These vents may serve as an example of how future oceans may look after acidification sets in—our oceans may be dominated by algae and invasives.

Ocean Acidification in the Great Barrier Reef

The Great Barrier Reef (GBR) is the largest continuous coral reef system in the world. Its economic, social, and icon assets are valued at AU$56 billion, owing to its vast biodiversity and services related to commercial and recreational fisheries, shoreline protection, and reef-related tourism and recreation. Ocean acidification poses a significant risk to these ecological and socioeconomic services, threatening not only the structural foundation of the GBR but the livelihoods of reef-dependent sectors of society. To assess the vulnerabilities of the GBR to ocean acidification, we review the characteristics of the GBR and the current valuation and factors affecting potential losses across three major areas of socioeconomic concern: fisheries, shoreline protection, and reef-related tourism and recreation. We then discuss potential solutions, both conventional and unconventional, for mitigating ocean acidification impacts on the GBR and propose a suite of actions that would help assess and increase the region's preparedness for the effects of ocean acidification.

The Great Barrier Reef (GBR) is the largest living structure in the world, covering an area of more than 344,000 km². Long and relatively narrow, the GBR extends 2300 km alongside Australia's northeast coast with its width ranging between 100 km in the north to 200 km in the south. The reef begins in the north at Australia's Cape York Peninsula and ends midway down the eastern coast at Lady Elliot Island, located just 90 km northeast of Bundaberg. 1,115,000 people live within the reef's catchment area that is made up of 35 river basins and, together with the GBR, totals 424,000 km² in area.

The GBR is the most famous and intensively managed marine park in the world. In 1975, the region gained protection through the creation of the Great Barrier Reef Marine Park (GBRMP) with a slightly larger area of 348,000 km² designated as a World Heritage Area (GBRWHA) in 1981, signifying the GBR's status as a place of global importance. Roughly 600 species of coral live on coral dominated reefs. These reefs, however, make up only about 7 percent of the GBRMP by area. Seagrass, mangroves, estuaries and other marine habitats help to host 100 species of jellyfish, 3000 varieties of molluscs, 500 species of worms, 1625 types of fish, 133 varieties of sharks and rays, more than 30 species of whales and dolphins, and various turtles and crocodiles.

Around the world, coral reefs are already under severe pressure from a number of stressors, including overfishing, pollution, and increasingly frequent and damaging bleaching events. Adding to this suite of threats, they are also among the most vulnerable ecosystems to ocean acidification (OA) because their very framework is dependent on calcium carbonate secreting organisms. Tropical coral reefs are identified as one of the most sensitive ecosystems in the Special Report on Global Warming of 1.5°C of the Intergovernmental Panel on Climate Change, with mass coral bleaching and mortality projected to increase due to interactions between rising ocean temperature, OA and increased frequency of storms. The report presents an extremely bleak outlook for these ecosystems, with a very high risk of loss of most (70%–90%) coral-dominated ecosystems and remaining structures being weakened due to OA if warming exceeds 1.5°C. The northern Great Barrier Reef already lost 50% of its shallow water corals during severe bleaching events in 2016–2017.

Coral reefs are biodiversity hotspots and provide habitat to a myriad of organisms, including many fish species. Loss of coral cover, whether due to OA, warming or other pressures on the reef, will lead to a shift in fish communities from species that prefer coral habitats toward species which are successful outside reef settings, with associated potential changes to important reef fisheries. Coral reefs also provide coastal protection from storms and support livelihoods and economic activities such as reef-associated tourism and recreation.

A recent valuation exercise strived to include the social and icon brand value of the Great Barrier Reef and found the total value of the reef to be AU$56 billion, owing to its vast biodiversity and assets related to commercial and recreational fisheries, shoreline protection, and reef-related tourism and recreation. This includes the support of 64,000 jobs in Australia. More recently, a social science approach was undertaken to identify the non-material value of the Great Barrier Reef to people. This approach assessed the importance of the GBR for providing lifestyle, sense of place, pride, identity, well-being, and aesthetic, scientific, and biodiversity values according to 8300 people across multiple cultural groups. People across all groups related strongly to all of the cultural values, highlighting the importance of non-material benefits that people derive from iconic ecosystems such as the GBR to people. Yet, these studies tend to oversimplify the value of the GBR and often fail to account for the ways in which a loss of coral reef resources in the GBR will affect the local and regional economies of Queensland or the rest of the world.

Impacts of Ocean Acidification on the Great Barrier Reef

Ocean acidification refers to the shifts in seawater chemistry that occur as a result of uptake of atmospheric carbon dioxide by the upper layers (300 m) of the ocean. When OA emerged as a dedicated research field in the late 1990s, corals and coral reefs were rapidly identified as potentially vulnerable given their role and sensitivity as key marine calcifiers, and several of the earliest studies on OA focused on corals and coral reefs. About 15% (579 papers out of 3648) of all papers published to date investigating a biological response to OA have looked at impacts on corals, which represents the third main taxonomic group studied after mollusks (674 papers) and phytoplankton (632 studies). The body of research on OA impacts on corals includes both laboratory and field studies, and many have been carried out in the real-world context of simultaneous warming and acidification. Although results are variable, the overall picture emerging from the research effort to date is that corals and coral reef systems are among the most vulnerable organisms and ecosystems to OA. This is in large part owing to the reliance of coral reefs on the capacity of corals and other calcifiers to produce calcium carbonate through the process of calcification, and existing calcareous structures' resistance to the process of dissolution, both of which are subject to negative impacts from changing carbonate chemistry conditions associated with OA. In addition to calcification, other potential processes susceptible to OA include reproduction, respiration, and photosynthesis, in both corals and other reef organisms such as algae and fish.

Several studies have looked specifically at GBR species and communities. A broad review of the implications of climate change, including OA, was compiled as part of the comprehensive climate vulnerability assessment for the GBR. This, in combination with more detailed studies published since, have shown a broad array of possible impacts on corals and coralline algae under future OA and warming, e.g. decreased calcification, primary production, settlement, reproduction, and survivorship, increased skeletal dissolution, and changes to gene expression, especially in early life stages. Webster found evidence for altered microbial communities in biofilms of a GBR crustose coralline algae, affecting its ability to perform its role as a substrate for coral settlement. OA and warming have been shown to accelerate bio-erosion of corals by microbial communities, endolithic algae and excavating sponges, adding to the corrosive effects of OA, although these organisms may themselves be susceptible to OA and warming which may limit the negative impacts they will cause in the future ocean. Several studies have been carried out on GBR fish, with results indicating a change in behavioral and sensory function such as attraction to predator scent, including in commercially important species such as the coral trout, although similar changes were not found in other coral reef fish species. Other examples of limited or positive impacts of OA on the GBR have been found, e.g. several GBR seagrass species seem to increase net photosynthesis rates under OA and biotic processes (e.g. photosynthesis) in reef sediments seem unaffected by OA. It is not yet fully understood if OA increases corals' susceptibility to bleaching even though it seems increasingly unlikely.

OA has the potential to affect not only biological processes but also ecological interactions between species, with some species benefitting to the detriment of others. For example, seaweed may become increasingly competitive compared with corals under future OA conditions on the GBR.

Coral and coralline algae communities present in naturally acidified waters around CO_2 seeps in Papua New Guinea are less diversified and complex as compared to similar communities outside the seep site. Less diverse and less structural complexity translate to less appropriate habitat for fish and other reef organisms with potential impacts on fisheries and other ecosystem services. Such studies provide 'windows into the future' and can, together with other methods, provide some much-needed insight into responses at the ecosystem level, necessary to understand any changes to services provided by those ecosystems.

It is likely that GBR communities already calcify less due to OA. Calcification rates increased by 25% in small patch reefs in mesocosm experiments when carbonate chemistry was restored to pre-industrial compared to present-day conditions and Albright found that net community calcification increased when water with conditions corresponding to preindustrial levels were applied to a GBR reef flat in a controlled field perturbation experiment. Decreased calcification is supported by results from skeletal records of massive corals from the inshore Great Barrier Reef, which indicate an 11% decline in calcification between 1990 and 2005, the fastest and most severe decline in at least 400 years. Another study argues that decreased community calcification on the Lizard Island reef flat over the last three decades might be primarily due to OA.

According to a review of regional accretion rates by Kennedy, the Great Barrier Reef has much lower net accretion rates when compared to areas such as the Coral Triangle, suggesting that the GBR may have a relatively higher sensitivity to OA in comparison to other reef systems. Dove showed that reefs may transition from net calcium carbonate accretion to net dissolution by the end of this century, which has also been confirmed in other areas of the world, at CO_2 seep sites, by CO_2 enrichment experiments in the field and by model projections.

Like in other reef systems, carbonate chemistry is highly variable on the GBR, both in time and space, and driven by both physical (e.g. temperature, mixing with water masses from adjacent waters) and biological (photosynthesis, respiration) processes. Corals are likely to experience changes in pH which go beyond declines projected for the end of the century on a regular basis. It is unknown though if this high natural variability confers enhanced resistance to OA. Many laboratory experiments to date have used scenarios of open ocean carbonate chemistry conditions rather than more locally relevant conditions. Cornwall found limited response and some evidence for faster calcification under extreme OA in corals and coralline algae from a site with high daily pH variability in North West Australia compared to a low-variability site. There seems to be little evidence for acclimation and adaptation to OA in the GBR.

Field and model studies converge to show that we can expect OA, particularly in combination with warming, to cause major changes in GBR communities, including loss of reef framework, biodiversity and ecosystem services. While warming remains the most acute concern for the GBR, with mass bleaching events expected to continue in the years to come, OA adds to the stress from warming and makes reefs less resilient, slowing recovery after bleaching events.

Potential Socio-economic Impacts of Great Barrier Reef Loss

While the evidence for adverse effects of OA and climate change on corals and coral reef ecosystems grows, and our capacity to project future changes improves, the challenge remains to project what these effects will mean for human communities depending on the reefs. The estimation of future losses in economic and societal value of coral reefs is complicated by the uncertainty associated with projections of human behavior in response to degradation of coral reefs, since human behavioral responses are notoriously difficult to predict with confidence given available data and knowledge of system dynamics.

Hoegh-Guldberg review existing literature on the potential economic consequences of losses to coral reef fisheries, coastal protection and tourism, and discuss factors affecting these losses (this issue). For instance, the authors point out that people may simply continue to take advantage of the decreased services provided by the reefs, albeit with less profit, enjoyment etc., or shift to using substitutes for lost services (e.g. recreation activities which are not dependent on the reef). These same reef users could also turn to other ecosystems that could provide similar services (e.g. mangroves in the case of shoreline protection and tourism), adapt their activities, or move. Regions like the GBR where most people do not rely on the reef as primary source of food, and where there are more options to adapt, would tend to be less vulnerable and more resilient to change.

Great Barrier Reef Fisheries

Current Value

The estimated economic contribution from the GBR fisheries cannot be wholly ascribed to the coral reef within the marine park because much of the park is habitat for non-reef species such as pelagic fish. Valuations that have examined the economic contribution of commercial and/or recreational fisheries within the GBR, with the exception of Teh have not distinguished between reef-dependent and non-reef dependent fisheries. Nevertheless, the range of estimates of the economic contribution from GBR fisheries provides insight into the value of reef ecosystems as a component of GBR fish habitats.

Many studies or reports that examined GBR fisheries estimated the gross "value" of the GBR's commercial fisheries – a measure of revenues. Gross revenues can be useful in determining the societal importance of an ecosystem service, but are an overestimate of the "value" of the good or service because the costs of production inputs, including environmental degradation or depletion of natural resource stocks, are not accounted for. The estimates for the annual gross revenue from GBR commercial fisheries ranged between AU\$119 million (does not include aquaculture) (1999–00) and US\$199 million (includes aquaculture) to US\$407 million for reef-dependent fisheries.

Gross value added ("GVA") focuses more on the additional value created by coral reef fisheries and is comprised of wages earned, profits and production taxes (less subsidies) that result from GBR fishing activity. Beginning in 2005, Access Economics generated a series of reports for the GBRM-PA that examined the "economic contribution" of the Great Barrier Reef in terms of "value added" or, in other words, the value of gross output (total revenue) minus the intermediate costs of producing the goods and services. Deloitte Access Economics found that the annual value added from commercial fishing and aquaculture in and around the Great Barrier Reef was AU\$162 million for Australia (2015–16) of which AU\$116 million was considered "direct value" or the economic contribution resulting from consumer transactions within the commercial fishing sector. About AU\$95 million of the total revenues (AU\$199 million) came from line, net, pot and trawl fishing, but the contribution of coral reefs to these economic contributions was not calculated.

Gross Operating Surplus ("GOS") is a measure of net value and is, in simple terms, the GVA minus employee compensation, minus taxes on production and plus subsidies received. Oxford Economics used a GOS estimate from GBR commercial fishery activity as a proxy measurement for "producer surplus", or the amount that a producer receives above the amount that the producer is willing to accept – a measure of net value. Applying a GOS/GVA ratio of .62 (from 2004 Queensland Regional Input–Output Tables) to an earlier 2006–07 Access Economic GVA estimate of AU$65.7 million per year (adjusted to 2009 AU$), the study found an estimated annual GOS of AU$41 million from the GBRMP-dependent commercial fisheries (which includes non-reef dependent fisheries).

Recreational fishing on the Great Barrier Reef is of the same order of magnitude of economic importance as commercial fishing. Annual gross revenue estimates for recreational fishing ranged from AU$108 million to AU$240 million. Deloitte Access Economics estimated the total annual "recreational" expenditures to be AU$415 million, mostly made up of "equipment" at AU$241 million, followed by "fishing" at AU$70 million, and also "boating", "sailing" and "visiting an island". The annual value added from "recreation" was AU$346 million for Australia, of which AU$206 was direct value added. Oxford Economics, using the same method employed for commercial fisheries, estimated the annual GOS ("producer surplus") associated with GBR recreational fisheries to be AU$8.6 million.

Consumers, in this case the recreational fishers, also enjoy benefits from coral reefs that are beyond what they spend to access reef areas, and over the years, economic valuation methods, such as the travel cost method and contingent valuation have been used to estimate the "consumer surplus". Oxford Economics found the estimated annual consumer surplus for GBRMP recreational fisheries to be AU$70.1 million (the average of two transferred values from previous studies that were derived from survey work and the travel cost method.

Factors Affecting Potential Losses Associated with GBR Fisheries

The annual net economic value from commercial fishery associated with the Great Barrier Reef is likely to be on the order of just over AU$40 million/year. This represents the maximum amount of economic net value from fishing that would be lost if fishers simply stop fishing and inputs and costs were saved and inputs used elsewhere. Recreational fishing generated just under AU$9 million/year in net value to producers, and recreational fishers were estimated to enjoy approximately AU$70 million annually in net benefits. The proportion of these benefits that depend on coral reefs is unclear. If recreational fishers turn to other activities on the water, expenditures associated with recreational fishing may not change much (expenditures on boating, sailing, and equipment make up more than 80% of recreation related expenditures associated with access to the GBR.) With business-as-usual as the measure of impact of coral reef loss, these estimates are likely to be overestimates of the true net economic cost of coral reef loss to fishing.

Shoreline Protection Provided by the Great Barrier Reef

Current Value

The GBR's patchy series of 2900 coral reefs, made up of both barrier and fringing (about 760 of the total), provide coastal protection from storms, waves and erosion to more than 316,000 coastal residents. We found only two studies that provided estimated values for the coastal protection provided by the GBR. Estimated a shoreline protection value of US$629 million (currency year not provided) per year for all of Australia's 49,000 km² of coral reef, and it appears that this value was based on "transferred" 2001 property values from a Hawaii coral reef valuation. Scaled to the GBRMP area and adjusted for inflation and exchange rates, this figure translates to a value of GBRMP coastal protection of about AU$438 million per year.

Oxford Economics used the replacement cost method to estimate coastal protection provided by the GBRMP. By taking the cost to construct erosion preventative revetment walls ($2300 per meter) in South Mission beach, Australia (about 15 km south of Cairns City) and applying this cost to the GBRMP reef length (2300 km), the study estimated a capital cost of AU$5.3 billion for GBRMP coastal protection. Note, that replacement cost estimates are often discouraged because unless replacement is or would be undertaken, there is no way of knowing whether replacement costs are significantly higher than actual value people place on the lost service. Since land along the GBR coast is used in a variety of ways and has varying vulnerabilities to storms, waves, and erosion, it is problematic to use a single replacement cost for one area of GBRMP in order to estimate the value of coastal protection for the whole coastline. There is a clear need for more data collection and better estimates of the value of shoreline protection here.

Figure: The Great Barrier Reef Marine Park, Queensland Australia.

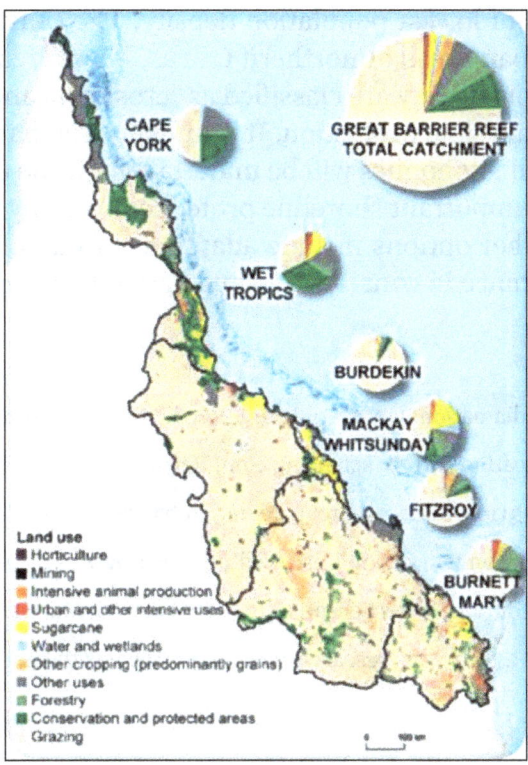

Figure: Land use with the Great Barrier Reef Marine Park.

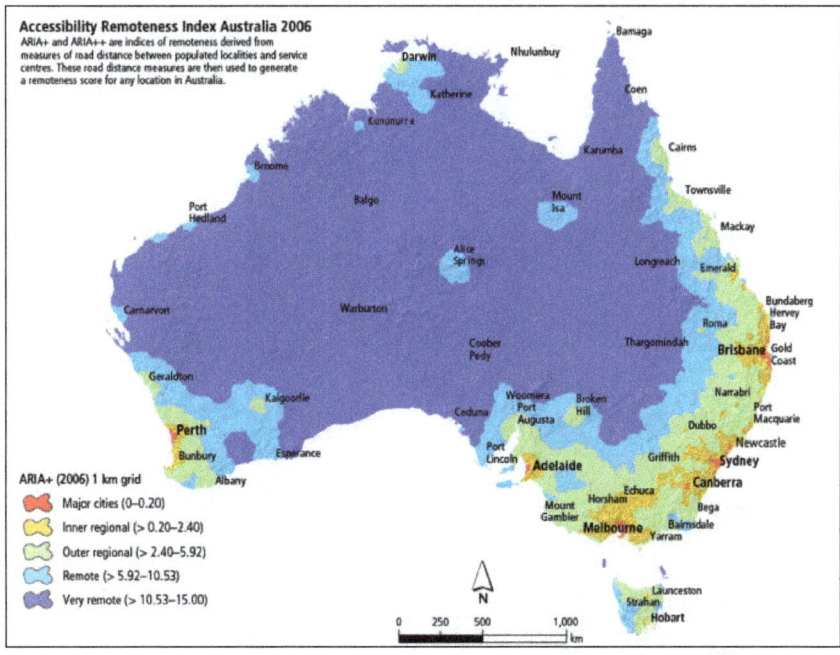

Figure: Accessibility remoteness Index for Australia.

Factors Affecting Potential Losses

Already, coastal areas within the Great Barrier Reef Catchment Area are subject to erosion. Only a small proportion of the coast protected by the Great Barrier Reef is developed for residential uses; much is considered outer-regional, remote, or very remote. The areas most affected by recent coral reef death also are the areas of lowest population density. The coastline in some developed and urban areas, especially in urban areas of northern Cairns, have already been hardened. Furthermore, many areas in the region already are classified as "erosion prone" and steps have been taken to address erosion and lost shoreline protection. It is unlikely that hard armoring of the shoreline, like that envisioned by Oxford Economics will be undertaken for the entire stretch of coast at risk. Mangroves may also provide important shoreline protection in many of the areas most affected by a loss of coral protection. Other options include adapting coastal structures to periodic flooding and managed retreat, for instance in conservation and agricultural areas.

References

- Marine-Ecosystems-1: media.nationalgeographic.org, Retrieved 22, June 2020

- Marine-ecosystem-classification-38170: sciencing.com, Retrieved 15, April 2020

- Marine-Plants-and-Algae-311671088: researchgate.net, Retrieved 09, August 2020

- Marine-chemistry: internetchemistry.com, Retrieved 13, May 2020

- Ocean-Acidification-The-Untold-Stories: oceana.org, Retrieved 06, March 2020

- The-Global-Oxygen-Cycle-277693427: researchgate.net, Retrieved 27, June 2020

Chapter 4

An Overview of the Ocean

Ocean refers to a continuous body of salt water that covers nearly three-fourth of the surface of the Earth. A few of the topics studied in relation to oceans are world's oceans and seas, ocean currents, ocean waves and ocean tides. The aim of this chapter is to explore the domain of oceans. These topics are crucial for a complete understanding of the subject.

World's Oceans and Seas

The ocean is a continuous body of salt water that covers more than 70 percent of the Earth's surface. Ocean currents govern the world's weather and churn a kaleidoscope of life. Humans depend on these teeming waters for comfort and survival, but global warming and overfishing threaten Earth's largest habitat.

Geographers divide the ocean into five major basins: the Pacific, Atlantic, Indian, Arctic, and Southern. Smaller ocean regions such as the Mediterranean Sea, Gulf of Mexico, and the Bay of Bengal are called seas, gulfs, and bays. Inland bodies of saltwater such as the Caspian Sea and the Great Salt Lake are distinct from the world's oceans.

The oceans hold about 321 million cubic miles (1.34 billion cubic kilometers) of water, which is roughly 97 percent of Earth's water supply. Seawater's weight is about 3.5 percent dissolved salt; oceans are also rich in chlorine, magnesium, and calcium. The oceans absorb the sun's heat, transferring it to the atmosphere and distributing it around the world. This conveyor belt of heat drives global weather patterns and helps regulate temperatures on land, acting as a heater in the winter and an air conditioner in the summer.

Figure: Surf and spray scatter as a large wave crashes onto the shore
in Palau. More than 250 islands make up the country.

Sea Life

The oceans are home to millions of Earth's plants and animals—from tiny single-celled organisms to the gargantuan blue whale, the planet's largest living animal. Fish, octopuses, squid, eels, dolphins, and whales swim the open waters while crabs, octopuses, starfish, oysters, and snails crawl and scoot along the ocean bottom.

Life in the ocean depends on phytoplankton, mostly microscopic organisms that float at the surface and, through photosynthesis, produce about half of the world's oxygen. Other fodder for sea dwellers includes seaweed and kelp, which are types of algae, and seagrasses, which grow in shallower areas where they can catch sunlight.

The deepest reaches of the ocean were once thought to be devoid of life, since no light penetrates beyond 1,000 meters (3,300 feet). But then hydrothermal vents were discovered. These chimney-like structures allow tube worms, clams, mussels, and other organisms to survive not via photosynthesis but chemosynthesis, in which microbes convert chemicals released by the vents into energy. Bizarre fish with sensitive eyes, translucent flesh, and bioluminescent lures jutting from their heads lurk about in nearby waters, often surviving by eating bits of organic waste and flesh that rain down from above, or on the animals that feed on those bits.

Despite regular discoveries about the ocean and its denizens, much remains unknown. More than 80 percent of the ocean is unmapped and unexplored, which leaves open the question of how many species there are yet to be discovered. At the same time, the ocean hosts some of the world's oldest creatures: Jellyfish have been around more than half a billion years, horseshoe crabs almost as long.

Other long-lived species are in crisis. The tiny, soft-bodied organisms known as coral, which form reefs mostly found in shallow tropical waters, are threatened by pollution, sedimentation, and global warming. Researchers are seeking ways to preserve fragile, ailing ecosystems such as Australia's Great Barrier Reef.

Human Impacts

Human activities affect nearly all parts of the ocean. Lost and discarded fishing nets continue to lethally snare fish, seabirds, and marine mammals as they drift. Ships spill oil and garbage; they also transport critters to alien habitats unprepared for their arrival, turning them into invasive species. Mangrove forests are cleared for homes and industry. Our garbage—particularly plastic—chokes the seas, creating vast "garbage patches" such as the Great Pacific Garbage Patch. Fertilizer runoff from farms turns vast swaths of the ocean into dead zones, including a New Jersey-size area in the Gulf of Mexico.

Climate change, the term scientists now use to describe global warming and other trends currently affecting the planet because of high greenhouse gas emissions from humans, is strikingly reflected in the oceans. The year 2018 marked the oceans' hottest year on record, and warmer waters lead to a range of consequences, from changing colors to rising sea levels to more frequent powerful storms. The greenhouse gas carbon dioxide is also turning ocean waters acidic, and an influx of freshwater from melting glaciers threatens to alter the weather-driving currents: the Atlantic Ocean's currents have slowed by about 15 percent over the past few decades.

A community of scientists, explorers, and citizen scientists continues to study the ocean, hoping that more information will yield more paths for conservation. Underwater drones, for example, are being deployed to explore undersea frontiers, while new tools are helping scientists measure and understand what they find.

Our blue planet is incredible. Stretching far and wide, huge bodies of saline water hug land masses and dominate the surface. Together, they cover around three quarters of the entire Earth's surface, and sink deep into a vastly unexplored abyss. Still much remains unknown, although much has been discovered, in our oceans. Through geo-mapping and political arrangement, there are officially five – each with their own biodiversity, topography and quirks.

Pacific Ocean

The Pacific Ocean is the largest ocean, around 15 times the size of the United States. With 25,000 islands in the region, the ocean also contains the most biodiverse waters in the world – thanks to the Coral Triangle. Even though the Pacific Ocean is best known for its incredible fauna, it also contains an incredible array of plants and coral reefs.

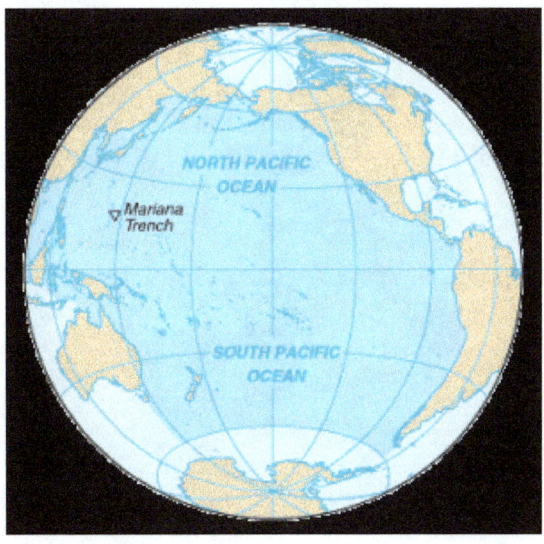

- Size: 165.25 million square kilometres.

- Average depth: 4,280 metres.

- Deepest Point: Mariana Trench, 10,911 metres.

- Surface temperature: From -1.4°C in poleward areas to 30°C near to the equator.

- Covers: 30.5% of Earth's total surface area.

- Boundaries: Asia, Australia, the Americas.

- Notable dive locations: Great Barrier Reef, Lembeh Strait, Komodo, Sipadan, Palau, Malapascua, Tubbataha, Chuuk Lagoon, Raja Ampat.

  ◦ 60% of the world's fish come from The Pacific Ocean.

- The Pacific Ocean shrinks in size by just over two centimetres each year.

- The Pacific Ocean Basin is home to 75% of the world's volcanoes.

- There are more than 25,000 islands in the Pacific.

- Pacific Ocean was declared "Mar Pacifico" in 1521 which is Portuguese for "Peaceful Sea".

Atlantic Ocean

Containing most of our planet's shallow seas – but with relatively few islands – the Atlantic Ocean is a relatively narrow body of water that snakes between nearly parallel continental masses, the Americas, Europe and Africa. It is famed for offering incredible encounters with large pelagics in the Caribbean, Mediterranean, Baltic, and the Gulf of Mexico. The Mid-Atlantic Ridge, that runs roughly down the centre of the ocean, separates the Atlantic Ocean into two large basins.

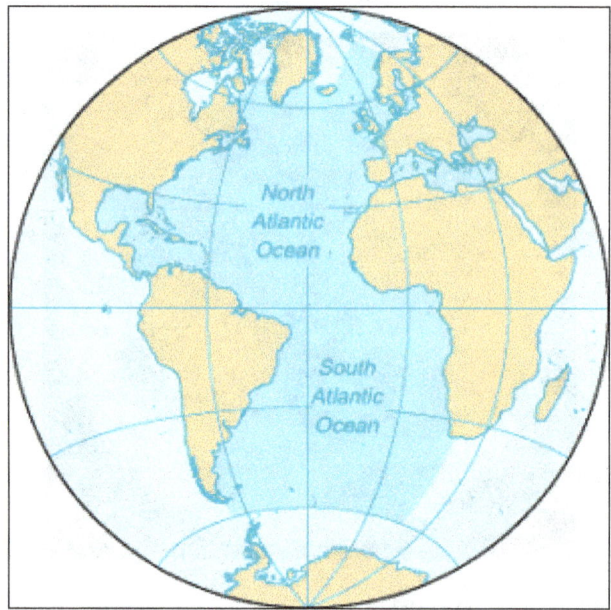

- Size: 82.36 million square kilometres.

- Average depth: 3,339 metres.

- Deepest point: Puerto Rico Trench, 8,605 metres.

- Surface temperature: From -2°C in the polar regions to over 30°C north of the equator.

- Covers: 20.8% of Earth's total surface area.

- Boundaries: The Americas, Europe, Africa and Antarctica.

- Notable dive locations: Grand Bahama, Cozumel, Grand Cayman, British Virgin Islands, Bermuda, Bonaire, The Great Blue Hole, Honduras.

- The name Atlantic comes from the Greek word Atlantikos which was known in the English language at the time, as the Sea of Atlas.

- The Atlantic Ocean is the world's saltiest sea with a water salinity level of between 33 – 37 parts per thousand.

- It's the world's youngest ocean, formed long after the Pacific, Indian and Arctic Oceans of the Triassic Period.

- Home to the earth's largest mountain range, The Mid Atlantic Ridge, which is 40,000 kilometres long by 1,601 kilometres wide – dividing the ocean into two distinct east and west regions.

- The Atlantic is famous for being the home of the legendary area known as the Bermuda Triangle, an area renowned for the mysterious disappearance of several aircraft and ships.

Indian Ocean

The Indian Ocean is enclosed on three sides by landmasses of Africa, Asia, and Australia. The southern border is wide open and exchanges with the much colder Southern Ocean. With relatively few islands, the continental shelf areas tend to be quite narrow and not many shallow seas exist. Some of the major rivers flowing into the Indian Ocean include the Zambezi, Indus, and the Ganges. Because much of the Indian Ocean lies within the tropics, this basin has the warmest surface ocean temperature.

- Size: 73.56 million square kilometres.

- Average depth: 3,960 metres.

- Deepest point: Sunda Deep, 7,450 metres.

- Surface temperature: N/A.

- Covers: 14.4% of Earth's total surface area.

- Boundaries: Africa, Asia, Australia/Oceania.

- Notable dive locations: Seychelles, Oman, Maldives, Musandam, Bali.

  - The ocean is the warmest ocean in the world and offers little scope to plankton and other species for growth.

  - It is estimated that approximately 40% of the world's oil comes from the Indian Ocean.

  - There was a discovery of a submerged continent in the Indian Ocean named the Kerguelen Plateau, it is believed to be of volcanic origins.

  - The Ocean's water evaporates at an abnormally high rate due to its temperature.

  - Every year it is estimated that the Indian Ocean becomes approximately 20 centimetres wider.

Southern Ocean

Compared to the other five oceans, the floor of the Southern Ocean is quite deep – ranging from 4,000 to 5,000 metres below sea level over most of the area that it occupies. In September of each year, a mobile icepack situated around the Antarctic reaches its greatest seasonal extent covering around 19 million square kilometres– later in March the icepack shrinks by almost 85%.

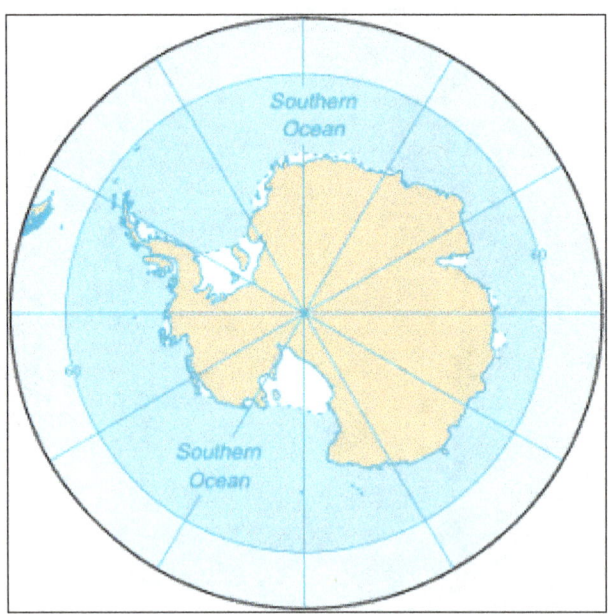

- Size: 20.3 million square kilometres.

- Average depth: 4,496 metres.

- Deepest point: South Sandwich Trench, 7,235 metres.

- Surface temperature: -2 to 10°C.

- Covers: 4.0% Earth's total surface area.

- Boundaries: Antarctica.

- Notable dive locations: Polar diving in Antarctica.

    ◦ The world's largest penguin species, the emperor penguin, lives on the ice of the Southern Ocean and on the Antarctica continent. Along with the world's largest animal, the blue whale, who often calls these waters home.

    ◦ Antarctica is home to 90% of the world's ice. This continent contained within the Southern Ocean's boundaries is the windiest, driest and coldest continent in the world.

    ◦ Having been only officially recognized in 2000, there is still some controversy as to whether it should be considered a separate ocean or merely an extension of the Atlantic, Pacific and Indian Oceans.

    ◦ The ocean is the youngest of the five oceans at only 30 million years of age and formed when the continents of South America and Antarctica completely split apart.

    ◦ Clouds are brighter in the Southern Ocean due to large plankton blooms, which release gases that allow water droplets to spread out more thus creating more reflective clouds.

Arctic Ocean

The world's smallest and shallowest (on average) ocean is also one of its most interesting. The crown of the world, both above and below the waves, enchanting creatures from narwhals to belugas sound out in the deep depths.

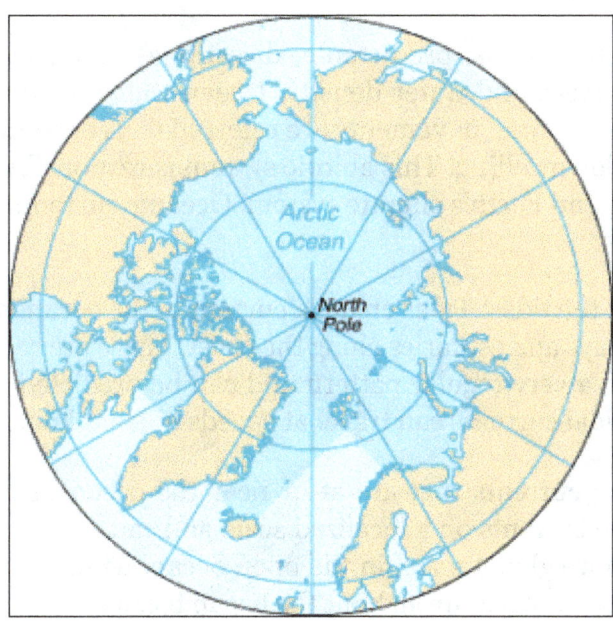

- Size: 14.05 million square kilometres.

- Average depth: 1,050 metres.

- Deepest Point: Litke Deep in the Eurasian Basin, 5,450 metres.

- Surface temperature: Average -1.8°C.

- Covers: 2.8% of Earth's total surface area.

- Boundaries: Europe, Asia, North America.

- Notable dive locations: Greenland, Baffin Island.

  - There are four whale species in the Arctic Ocean including the bowhead whale, grey whale, narwhal, and beluga whale.

  - When the ice of the Arctic Ocean melts it releases nutrients and organisms into the water which promotes the growth of algae. The algae feed zooplankton which serves as food for the sea life.

  - Because of the Arctic Ocean's low evaporation, large freshwater inflow, and its limited connection to other oceans it has the lowest salinity of all oceans. Its salinity varies depending on the ice covers' freezing and melting.

  - Icebergs often form or break away from glaciers posing a threat to ships the most famous being the Titanic. Ships also often get trapped or crushed by the ice.

  - Ice cover of the ocean is shrinking due to global warming, and it has been observed that the rate of disappearance of ice cover is 3% per decade.

Ocean currents are the continuous, predictable, directional movement of seawater driven by gravity, wind (Coriolis Effect), and water density. Ocean water moves in two directions: horizontally and vertically. Horizontal movements are referred to as currents, while vertical changes are called upwellings or downwelling. This abiotic system is responsible for the transfer of heat, variations in biodiversity, and Earth's climate system. Oceanic currents are driven by three main factors:

- The rise and fall of the tides: Tides create a current in the oceans, which are strongest near the shore, and in bays and estuaries along the coast. These are called "tidal currents." Tidal currents change in a very regular pattern and can be predicted for future dates. In some locations, strong tidal currents can travel at speeds of eight knots or more.

- Wind: Winds drive currents that are at or near the ocean's surface. Near coastal areas winds tend to drive currents on a localized scale and can result in phenomena like coastal upwelling. On a more global scale, in the open ocean, winds drive currents that circulate water for thousands of miles throughout the ocean basins.

- Thermohaline circulation: This is a process driven by density differences in water due to temperature (thermo) and salinity (haline) variations in different parts of the ocean. Currents driven by thermohaline circulation occur at both deep and shallow ocean levels and move much slower than tidal or surface currents.

Types of Ocean Currents

Surface Currents

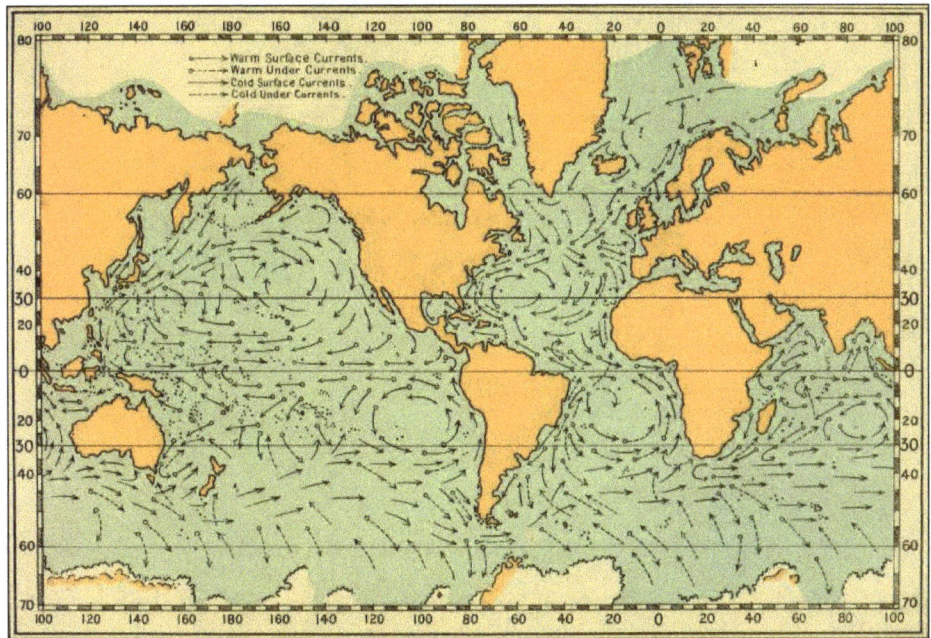

Figure: This is a map ocean surface currents from 1877.

At the surface, currents are mainly driven by four factors—wind, the Sun's radiation, gravity, and Earth's rotation. All of these factors are interconnected. The Sun's radiation creates prevailing wind patterns, which push ocean water to bunch in hills and valleys. Gravity pulls the water away from hills and toward valleys and Earth's rotation steers the moving water.

Sun and Wind

Wind is a major force in propelling water across the globe in surface currents. When air moves across the ocean's surface, it pulls the top layers of water with it through friction, the force of resistance between two touching materials moving over one another. Surface ocean currents are driven by consistent wind patterns that persist throughout time over the entire globe, such as the jet stream. These wind patterns (convection cells) are created by radiation from the Sun beating down on Earth and generating heat.

The Sun's radiation is strongest at the equator and dissipates the closer you get to the poles. This uneven distribution of heat causes air to move. The hot air over the equator rises and moves away from the equator. Likewise, cold air from the poles sinks and moves towards the equator. The clashing of hot air originating at the equator and cold air originating at the poles creates regions of high atmospheric pressure and low atmospheric pressure along specific latitude lines. It would make intuitive sense that the hot air and cool air would meet in the middle of the equator and the North or South Pole, however, in reality it is much more complicated. A combination of Earth's rotation, the fact that Earth is tilted on an axis, and the placement of most continents in the Northern Hemisphere, create pressure systems that divide each hemisphere into three distinct wind patterns or circulation cells.

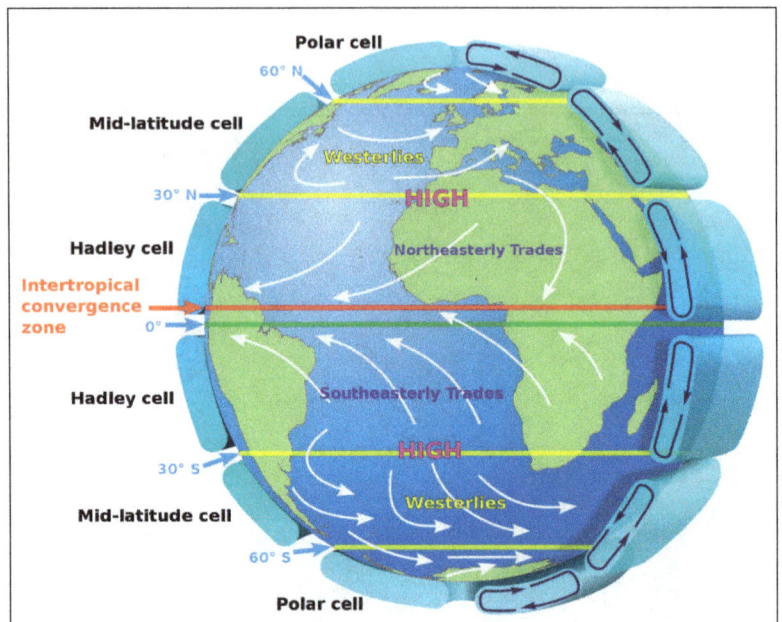

Figure: The major wind patterns drive oceanic currents.

In the Northern Hemisphere, the most northern system, the polar cell, blows air in a consistent southwestern direction toward a pocket of low pressure along the 60-degree latitude line. The middle system, the Ferrel cell, blows in a consistent northeastern direction toward the same 60-degree low. And the most southern system, the Hadley cell, blows air in a consistent southwestern direction toward a region of low pressure along the equator. The result is a global pattern of prevailing wind, and it is this consistent wind that impacts the ocean.

While it may appear that the ocean is a flat surface, the reality is that it is a series of hills and valleys in the water. At the places where the wind generated currents converge into each other, the ocean water is pushed to build a slight hill. Likewise, where the winds diverge, the ocean water dips in a slight depression.

Gravity and Earth's Rotation

Wind pushes water into hills of high pressure which leave behind valleys of low pressure. Since water is a liquid that prefers to stay at a level height, this creates an unstable situation. Following the pull of gravity, ocean water moves from the built-up areas of high pressure down to the valleys of low pressure. But as the water moves from hills to valleys, it does so in a curved trajectory, not a straight line. This curving is a result of Earth's spin on its axis.

On Earth, movement in a straight line over long distances is harder than it may seem. That's because Earth is constantly rotating, meaning every object on its surface is moving at the speed at which the Earth is spinning on its axis. From our perspective, stationary objects are just that unmoving. In reality, they are whipping around at a speed of roughly 1,000 miles per hour (1600 km/hr) at Earth's equator. It is that whipping, rotating motion that influences the movement of any object not in direct contact with the planet's surface, making straight appearing trajectories actually bend. It also influences the movement of ocean currents. Scientists refer to this bending as the Coriolis Effect.

It is easiest to understand this phenomenon when thinking about travel in a northern or southern direction. Since Earth is essentially a sphere and it spins around an axis, anything near Earth's equator will travel the fastest—since Earth is rotating at a constant rate and the equator runs along the widest part of the sphere, any object there must travel the entirety of Earth's circumference in one rotation. As you get closer and closer to the poles, the distance traveled in one rotation gradually shrinks until it reaches zero at either pole. Therefore, an object on the surface will gradually spin slower the closer it gets to a pole.

But leave the surface of the planet, and the anchor keeping you in sync with the land beneath you disappears. Any moving object (plane, boat, hot air balloon, water) will begin its travels at the rotating speed of the location where it took off from. If it should travel north or south, the ground beneath it will be traveling at a different speed. Travel North from the Equator, and the ground will gradually spin slower beneath you. This causes an object attempting to travel in a straight line to veer to the right in the Northern Hemisphere and veer to the left in the Southern Hemisphere relative to the direction traveling.

Understanding how the rotating Earth affects movement to the west or east is a bit trickier. Envision an elastic string attached to a ball on one end and an anchored point at the other. The faster the ball is spun around the anchor, the more the elastic stretches and the farther the ball travels from the center point. An object traveling on Earth behaves the same way. If the object moves east, in the direction that Earth is spinning, it is now traveling around the axis of Earth faster than it was when it was anchored—and so, the object wants to move out and away from the axis. Still tethered by gravity, the object does so by moving toward the equator, the place on Earth that is the greatest distance from the axis. Travel west, the opposite direction that Earth is spinning, and now the object is spinning slower than Earth's surface and so it wants to move toward the axis. It does so by moving toward the pole. This again appears as a bend to the right in the Northern hemisphere and to the left in the Southern hemisphere.

Water moving along Earth's surface is also subject to the Coriolis effect which causes moving water to curve in the same directions described above. In the Northern Hemisphere, surface water curves to the right and in the Southern Hemisphere it curves to the left of the direction it is forced to move.

Swirling Gyres

Earth's rotation is also responsible for the circular motion of ocean currents. There are 5 major gyres—expansive currents that span entire oceans—on Earth. There are gyres in the Northern Atlantic, the Southern Atlantic, the Northern Pacific, the Southern Pacific, and the Indian Ocean. Similar to surface waters, Northern gyres spin clockwise (to the right) while gyres in the south spin counterclockwise (to the left).

The center of the gyres is relatively calm areas of the ocean. The Sargasso Sea, known for its vast expanses of floating Sargassum seaweed, exists in the North Atlantic gyre and is the only sea without land boundaries. Today, gyres are also areas where marine plastic and debris congregate. The most famous one is known as the Great Pacific Garbage Patch, but all five gyres are centers of plastic accumulation.

Ekman Transport

Wind moving across the ocean moves the water beneath it, but not in the way you might expect. The Coriolis Effect, the apparent force created by the spinning of Earth on its axis, affects water movement, including movement instigated by wind. Recall that Coriolis causes the trajectory of a moving object to veer to the right or the left depending upon the hemisphere it is located in. But in this case, the three-dimensional nature of the ocean plays into the direction of the water's overall movement. Wind blowing over water will move the ocean water underneath it in an average direction perpendicular to the direction the wind is traveling.

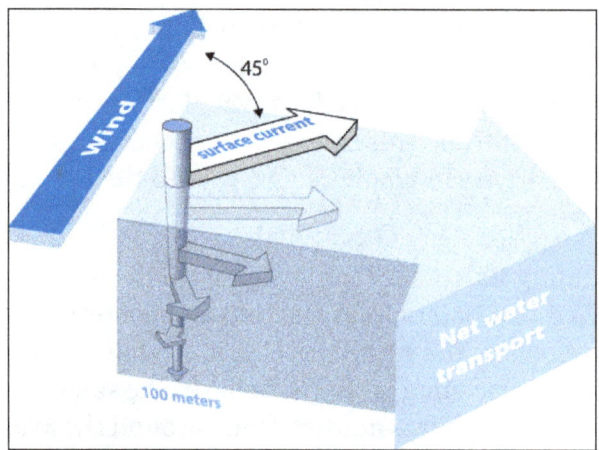

Figure: Ekman transport creates a spiral as wind drags the surface
of the ocean, which then drags deeper layers of water.

As wind blows over the surface layer of water, friction between the two pulls the water forward. As we know, when water (and other objects) moves across Earth's surface it bends due to the Coriolis Effect. The top most layer of water will bend away from the direction of the wind at about 45 degrees. For simplicity, we will assume that this scenario is in the Northern Hemisphere and all movement bends to the right. As the top layer of water begins to travel, it in turn pulls on the water layer beneath it, just as the wind had. Now this second water layer begins to move, and it travels in a direction slightly to the right of the layer above it. This effect continues layer by layer as you move down from the surface, creating a spiral effect in the moving water.

In addition to a change in direction, each sequential layer down loses energy and moves at a slower speed. Friction causes the water to move, but drag resists that movement, so as we travel from the top layer to the next, some of the energy is lost. When all the layers down the spiral are accounted for, the net direction of the water is perpendicular to the direction of the wind.

Deep Currents

The ocean is connected by a massive circulatory current deep underwater. This planetary current pattern, called the global conveyor belt, slowly moves water around the world—taking 1,000 years to make a complete circuit. It is driven by changes in water temperature and salinity, a characteristic that has scientists refer to the current as an example of thermohaline circulation.

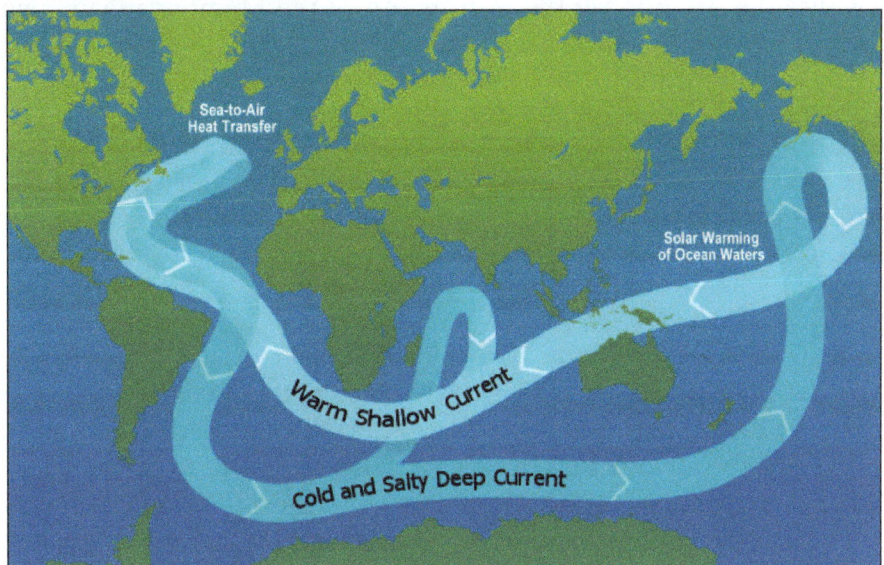

Figure: Differences in temperature and salinity drive deep ocean currents.

Both heat and salt contribute to the ocean water's density. Saltier and colder water is heavier and denser than less salty (or fresher), warmer water. Around the globe there are areas where the heat and saltiness of ocean water (and therefore, its density) change. The most important of these areas is in the North Atlantic.

As warm Atlantic water from the Equator reaches the cold polar region in the North via the Gulf Stream, it rapidly cools. This region is also cold enough that the ocean water freezes, but only the water turns to ice. As the water freezes it leaves the salt behind, causing the surrounding water to become saltier and saltier. The cold, salty water then sinks in a mass movement to the deep ocean. It is this sinking that is a main driver for the entire deep-water circulation system that moves massive quantities of water around the globe. Cooling also occurs near Antarctica, but not to the extremes that happen in the Northern Hemisphere.

Figure: Locals in Zebbug, Malta created salt pans where they can collect sea salt after the super salty Mediterranean Sea water has dried.

Another area of the ocean where massive amounts of water move to the ocean's depths is in the Mediterranean. In this area, evaporation is the main driver that changes the salinity of the ocean water. As water in the Mediterranean evaporates, it leaves the salt behind. This super salty ocean water then bleeds into the Atlantic via the thin mouth of the Mediterranean, also known as the Strait of Gibraltar.

When cold, salty water circulates the globe and gradually becomes warmer, it begins to rise. The "old" deep water is full of nutrients that have accumulated from the sinking of waste from the productive surface waters up above. Locations where the "old" water rises are highly productive areas because they contain ample nutrients and have access to sunlight—the perfect combination for photosynthesis.

Currents and Change

Because ocean circulation is driven by temperature change, any variation to the planet's climate could significantly alter the system. Scientists worry that the melting ice caused by global warming may weaken the global conveyer belt by adding extra fresh water in the Arctic. A 2018 study found that the massive ocean current that courses around the Atlantic Ocean, called the Atlantic Meridional Overturning Circulation, has decreased in strength by about 15 percent since 400 AD and is now the weakest it has been in 1,600 years. Ironically, despite an overall increase in global temperatures, many places in North America and Europe may get colder as a result.

Rip Currents

Figure: A rip current can be seen from up above.

Not all currents occur at such a large scale. Individual beaches may have rip currents that are dangerous to swimmers. Rip currents are strong, narrow, seaward flows of water that extend from close to the shoreline to outside of the surf zone. They are found on almost any beach with breaking waves and act as "rivers of the sea," moving sand, marine organisms, and other material offshore. Rip currents are formed when there are alongshore variations in wave breaking. In particular, rip currents tend to form in regions with less wave breaking sandwiched between regions of greater wave breaking. This can occur when there are gaps in sand bars nearshore, from structures like piers or jetties, or from natural variations in how waves are breaking.

Rip currents can move faster than an Olympic swimmer can swim, at speeds as fast as eight feet (2.4 meters) per second. At these speeds, a rip current can easily overpower a swimmer trying to return to shore. Instead of attempting to swim against the current, experts suggest not to fight it and to swim parallel to shore.

Currents and Nature

Unseen by the human eye, thousands of microscopic animals hitch rides across oceans on an oceanic highway. These animals, called zooplankton, move at the whim of ocean currents. Off the Eastern Shore of the United States, one of the most powerful ocean currents—the Gulf Stream—is transporting zooplankton from the Gulf of Mexico, around the tip of Florida, up to Cape Cod in Massachusetts and then across the North Atlantic Ocean towards Europe. The currents enable the young creatures to find their way to hospitable places where they grow into adults.

Figure: Currents on the ocean surface are driven by wind, temperature, gravity, and the spin of Earth on its axis.

Other ocean creatures hitch rides on currents using floating debris, like mats of seaweed, tree trunks, and even plastic. They use these havens to survive the otherwise perilous open ocean. After the 2011 tsunami that prompted the Fukushima Daiichi power plant meltdown in Japan, debris from the Japanese coast began washing ashore on the West coast of North America, bringing with it over 280 Japanese species. The movement of species across ocean basins helps maintain populations across the entirety of a species' range. It also ensures the diversity of genetics within a population, an important factor for keeping species resistant and resilient to hardships like disease and environmental disasters.

Figure: The Bay of Fundy in New Brunswick, Canada has the highest tidal range. The tides range from 3.5m (11ft) to 16m (53ft) and cause erosion to the landscape, creating massive cliffs.

Meanwhile, Earth continues to spin. As Earth rotates, the water bulges stay in line with the Moon whiles the planet's surface moves underneath it. A specific point on the planet will pass through both of the bulges and both of the valleys. When a specific place is in the location of a bulge it experiences a high tide. When a specific place is in the location of a valley it experiences a low tide. During one planetary rotation (or one day) a specific location will pass through both bulges and both valleys, and this is why we have two high tides and two low tides in a day. But, while Earth takes 24 hours to complete one rotation, it must then rotate an additional and 50 minutes to catch up with the orbiting Moon. This is why the time of high tide and the time of low tide change slightly every day.

Figure: As the tide recedes moored boats are left to sit in the muddy sand.

The Sun also has a part to play in causing the tides, and its location in relation to the Moon alters the strength of the pull on the ocean. When the Sun and Moon are in line with one another they reinforce each other's gravitational pulls and create larger-than-normal tides called spring tides. This happens when the Moon is either on the same side of Earth as the Sun or directly on the opposite side of Earth. Smaller-than-usual tidal ranges, called neap tides, occur when the gravitational force of the Sun is at a right angle to the pull from the Moon. The two forces of the Sun and Moon cancel each other out and create a neap tide.

Continental Interference

If Earth were a sphere covered by water, only the water would be able to move freely over the planet's surface and the two tides in a day at each location would be more or less the same. But continents obstruct the flow of water, causing this seemingly simple daily cycle to be a bit more complicated. Because of continental obstruction, some locations experience two tides a day that are more or less the same height (known as semidiurnal tides), some locations experience one tide at one height and the second at a different height (mixed semidiurnal tides), and some locations have so much interference from land that they only experience one high tide and one low tide per day (diurnal tides).

The local geography can also affect the way the tides behave in a location. Shores around coastal islands and inlets may experience delayed tides compared to smoother surrounding coasts since the water must funnel in through constrained waterways.

Tides and Nature

The intertidal zone, the coastal area tides submerge for part of the day, is home to many ocean creatures. It takes a special set of adaptations to live a life half the time scorched by the Sun and the other submerged underwater. Moreover, the incoming tide promises a constant pounding by ocean waves. Despite this, it's a place where species thrive. Shelled mollusks like periwinkles, muscles, and barnacles cling to rocks, sea stars wedge themselves in crevices, and crabs hide in fronds of algae.

Red Tide

Figure: Harmful algal blooms are dangerous, producing toxins that can kill marine organisms, taint shellfish, cause skin irritations, and even foul the air.

A red tide is not a true tide at all but rather a term used to describe the red color of an algal bloom. Algae are integral to ocean systems, but when they are supplied with excessive amounts of nutrients they can explode in number and smother other organisms. The algae may produce toxins or they can die, decay, and the bacteria decomposing them take up all the oxygen. This massive growth of algae can become harmful to both the environment and humans, which is why scientists often refer to them as harmful algal blooms or HABs.

Monitoring Tides

Tidal movements are tracked using networks of nearshore water level gauges, and many countries provide real-time information with tidal listings and tidal charts. Tides can be tracked at specific locations in order to predict the height of a tide, i.e. when low and high tide will occur in the future. The Bay of Fundy in Nova Scotia, Canada has the highest tidal range of any place on the planet. The tides there range from 11 feet (3.5 m) to 53 feet (16 m) and cause erosion, creating massive cliffs. This erosion also releases nutrients into the water that help support marine life. The currents associated with the tides are called flood currents (incoming tide) and ebb currents (outgoing tide). Having reliable knowledge about the tides and tidal currents is important for navigating ships safely, and for engineering projects such as tidal and wave energy, as well as for planning trips to the seashore.

Ocean Waves

Ocean waves are disturbances in the surface of the ocean. Ocean waves come in many shapes and sizes, ranging in length from a fraction of a centimeter for the smallest ripples to half the circumference of Earth for the tides. They are formed by wind, gravity, earthquakes, and submarine landslides disturbing the water surface. Once formed, and regardless of origin, ocean waves can travel great distances before reaching the coast. The ocean waves arriving at the shore today may have had their beginnings many hours or even days earlier a hemisphere away.

How are Ocean Waves Measured?

Ocean altimeter satellite missions, such as TOPEX/Poseidon and the Jason-series, measure significant wave height, which is the average wave height (from trough to crest) of the highest third of waves in a given sample period. The spacecrafts' radar altimeters measure the precise distance between the satellite and sea surface. The round-trip travel time of microwave pulses bounced from the spacecraft to the sea surface and back to the spacecraft provides data indicating sea surface height and the topography of the ocean surface. The precise altitude of the satellite is determined by a sophisticated estimation procedure based on instrument systems onboard the satellite and a network of ground receivers across the globe.

Causes of Ocean Waves

Looking out at the ocean, one often sees a seemingly infinite series of waves, transporting water from one place to the next. Though waves do cause the surface water to move, the idea that waves are travelling bodies of water is misleading.

Waves are actually energy passing through the water, causing it to move in a circular motion. When a wave encounters a surface object, the object appears to lurch forward and upward with the wave, but then falls down and back in an orbital rotation as the wave continues by, ending up in the same position as before the wave came by. If one imagines wave water itself following this same pattern, it is easier to understand ocean waves as simply the outward manifestation of kinetic energy propagating through seawater. In reality, the water in waves doesn't travel much at all. The only thing waves do transmit across the sea is energy.

The idea of waves being energy movement rather than water movement makes sense in the open ocean, but what about on the coast, where waves are clearly seen crashing dramatically onto shore? This phenomenon is a result of the wave's orbital motion being disturbed by the seafloor. As a wave passes through water, not only does the surface water follow an orbital motion, but a column of water below it (down to half of the wave's wavelength) completes the same movement. The approach of the bottom in shallow areas causes the lower portion of the wave to slow down and compress, forcing the wave's crest higher in the air. Eventually this imbalance in the wave reaches a breaking point, and the crest comes crashing down as wave energy is dissipated into the surf.

Where does a wave's energy come from? There are a few types of ocean waves and they are generally classified by the energy source that creates them. Most common are surface waves, caused by wind blowing along the air-water interface, creating a disturbance that steadily builds as wind continues to blow and the wave crest rises. Surface waves occur constantly all over the globe, and are the waves you see at the beach under normal conditions.

Adverse weather or natural events often produce larger and potentially hazardous waves. Severe storms moving inland often create a storm surge, a long wave caused by high winds and a continued low pressure area. Submarine earthquakes or landslides can displace a large amount of water very quickly, creating a series of very long waves called tsunamis. Storm surges and tsunamis do not create a typical crashing wave but rather a massive rise in sea level upon reaching shore, and they can be extremely destructive to coastal environments.

Anatomy of a Wave

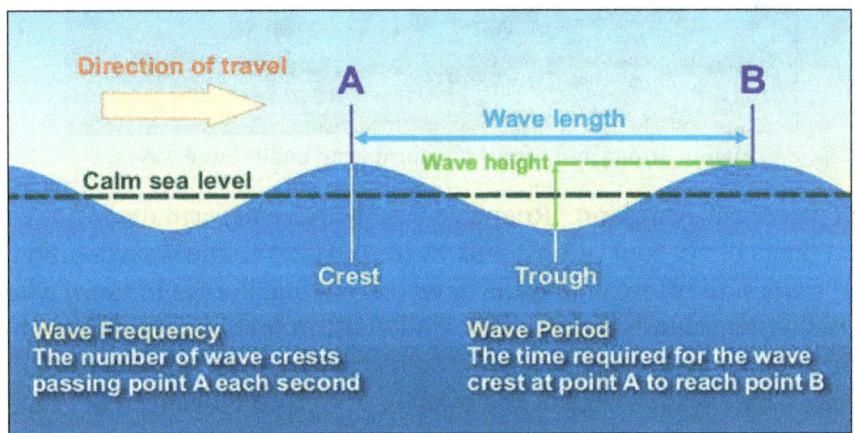

Figure: The anatomy of a wave.

A wave forms in a series of crests and troughs. The crests are the peak heights of the wave and the troughs are the lowest valleys. A wave is described by its wavelength (or the distance between two sequential crests or two sequential troughs), the wave period (or the time it takes a wave to travel the wavelength), and the wave frequency (the number of wave crests that pass by a fixed location in a given amount of time). When a wave travels, it is passing through the water, but the water barely travels, rather it moves in a circular motion.

Wave Formation

Surface Waves

Waves on the ocean surface are usually formed by wind. When wind blows, it transfers the energy through friction. The faster the wind, the longer it blows, or the farther it can blow uninterrupted, the bigger the waves. Therefore, a wave's size depends on wind speed, wind duration, and the area over which the wind is blowing (the fetch). This variability leads to waves of all shapes and sizes. The smallest categories of waves are ripples, growing less than one foot (.3 m) high. The largest waves occur where there are big expanses of open water that wind can affect. Places famous for big waves include Waimea Bay in Hawaii, Jaws in Maui, Mavericks in California, Mullaghmore Head in Ireland, and Teahupoo in Tahiti. These large wave sites attract surfers, although occasionally, waves get just too big to surf. Some of the biggest waves are generated by storms like hurricanes. In 2004, Hurricane Ivan created waves that averaged around 60 feet (18 meters) high and the largest were almost 100 feet (30.5 meters) high. In 2019, hurricane Dorian also created a wave over 100 feet high in the northern Atlantic.

Figure: Strong and persistent storm wind builds large waves.

Giant waves don't just occur near land. 'Rogue waves,' which can form during storms, are especially big—there are reports of 112 foot (34 m) and 70 foot (21 m) rogue waves—and can be extremely unpredictable. To sailors, they look like walls of water. No one knows for sure what causes a rogue wave to appear, but some scientists think that they tend to form when different ocean swells reinforce one another. Many of the largest rogue waves recorded have been in the North Sea in the North Atlantic Ocean. One was recorded by a buoy in 2013 and measured 62.3 feet (19 m) and another nicknamed the Draupner wave was a massive wall of water 84 feet (25.6 m) high that crossed a natural gas platform on New Year's Eve, 1995.

Tsunami Waves

Figure: A tsunami is a set of waves created by a disturbance, likely an earthquake, which reaches the surface of the sea.

A classic tsunami wave occurs when the tectonic plates beneath the ocean slip during an earthquake. The physical shift of the plates force water up and above the average sea level by a few meters. This then gets transferred into horizontal energy across the ocean's surface. From a single tectonic plate slip, waves radiate outwards in all directions moving away from the earthquake.

When a tsunami reaches shore, it begins to slow dramatically from contact with the bottom of the seafloor. As the leading part of the wave begins to slow, the remaining wave piles up behind it, causing the height of the wave to increase. Though tsunami waves are only a few feet to several meters high as they travel over the deep ocean, it is their speed and long wavelength that cause the change to dramatic heights when they are forced to slow at the shore.

Tsunami waves are capable of destroying seaside communities with wave heights that sometimes surpass around 66ft (20 m). Tsunamis have caused over 420,000 deaths since 1850—over 230,000 people were killed by the giant earthquake off Indonesia in 2004, and the damage caused to the Fukushima nuclear reactor in Japan by a tsunami in 2011 continues to wreak havoc. Although tsunamis cannot be predicted in advance when an earthquake occurs, tsunami warnings are broadcast and any waves can be tracked by a global network of buoys – this early warning system is essential because tsunamis can travel at over 400 miles per hour (644 km/hr.). The highest tsunami wave reached about 1,720 ft (524 m), a product of a massive earthquake and rockslide. When the wave hit shore, it was said to destroy everything.

There are also other, usually less destructive tsunami waves caused by weather systems called meteotsunamis. These tsunami waves have similar characteristics to the classical earthquake driven tsunamis described above, however they are typically much smaller and focused along smaller regions of the oceans or even Great Lakes. Meteotsunamis are often caused by fast moving storm systems and have been measured in several cases at over 6 feet (2 meters) high. A 2019 study found that smaller meteotsunami waves strike the east coast of the U.S. more than twenty times a year.

Ocean Tides

An ocean tide refers to the cyclic rise and fall of seawater. Tides are caused by slight variations in gravitational attraction between the Earth and the moon and the Sun in geometric relationship with locations on the Earth's surface. Tides are periodic primarily because of the cyclical influence of the Earth's rotation.

The moon is the primary factor controlling the temporal rhythm and height of tides. The moon produces two tidal bulges somewhere on the Earth through the effects of gravitational attraction. The height of these tidal bulges is controlled by the moon's gravitational force and the Earth's gravity pulling the water back toward the Earth. At the location on the Earth closest to the moon, seawater is drawn toward the moon because of the greater strength of gravitational attraction. On the opposite side of the Earth, another tidal bulge is produced away from the moon. However, this bulge is due to the fact that at this point on the Earth the force of the moon's gravity is at its weakest. Considering this information, any given point on the Earth's surface should experience two tidal crests and two tidal troughs during each tidal period.

Figure: The moon's gravitational pull is the primary force responsible for the tides on the Earth. Photo taken by the Galileo spacecraft from a distance of about 6.2 million kilometers from Earth, on December 16, 1992.

The timing of tidal events is related to the Earth's rotation and the revolution of the moon around the Earth. If the moon was stationary in space, the tidal cycle would be 24 hours long. However, the moon is in motion revolving around the Earth. One revolution takes about 27 days and adds about 50 minutes to the tidal cycle. As a result, the tidal period is 24 hours and 50 minutes in length.

The second factor controlling tides on the Earth's surface is the Sun's gravity. The height of the average solar tide is about 50% the average lunar tide. At certain times during the moon's revolution around the Earth, the direction of its gravitational attraction is aligned with the Sun's. During these times the two tide producing bodies act together to create the highest and lowest tides of the year. These spring tides occur every 14-15 days during full and new moons.

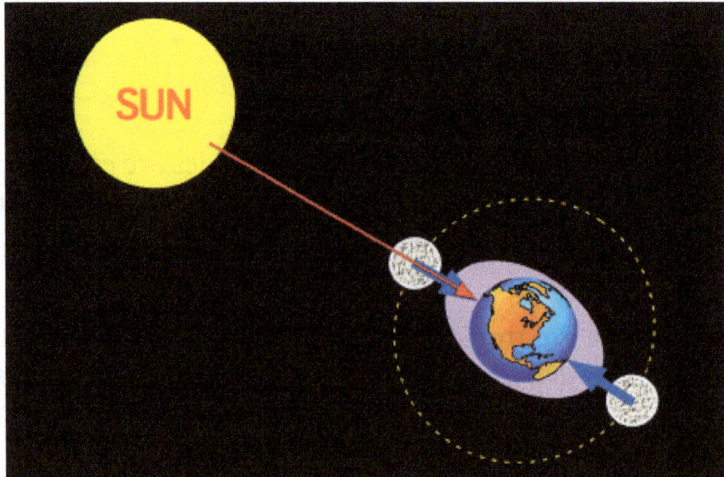

Figure: Forces involved in the formation of a spring tide.

When the gravitational pull of the moon and Sun are at right angles to each other, the daily tidal variations on the Earth are at their least. These events are called neap tides and they occur during the first and last quarter of the moon.

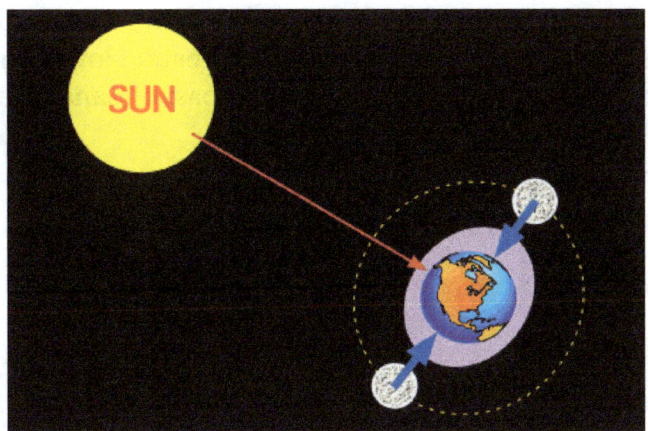

Figure: Forces involved in the formation of a neap tide.

Types of Tides

The geometric relationship of moon and Sun to locations on the Earth's surface results in creation of three different types of tides. In parts of the northern Gulf of Mexico and Southeast Asia, tides have one high and one low water per tidal day. These tides are called diurnal tides.

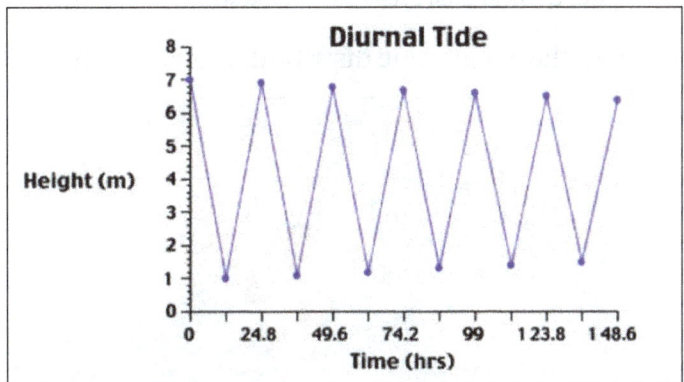

Figure: Cyclical tidal cycles associated with a diurnal tide.

Semi-diurnal tides have two high and two low waters per tidal day. They are common on the Atlantic coasts of the United States and Europe.

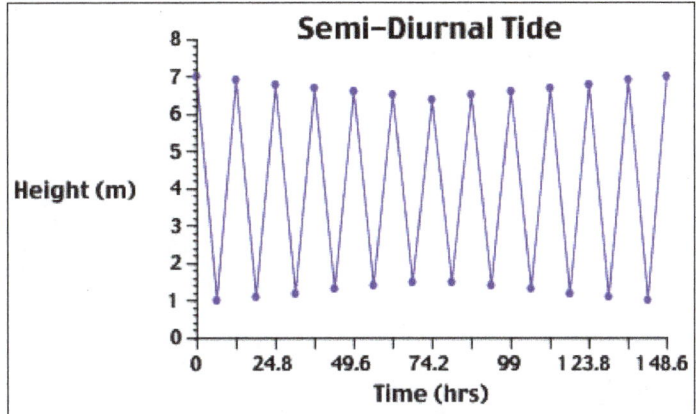

Figure: Cyclical tidal cycles associated with a semi-diurnal tide.

Many parts of the world experience mixed tides where successive high-water and low-water stands differ appreciably. In these tides, we have a higher high water and lower high water as well as higher low water and lower low water. The tides around west coast of Canada and the United States are of this type.

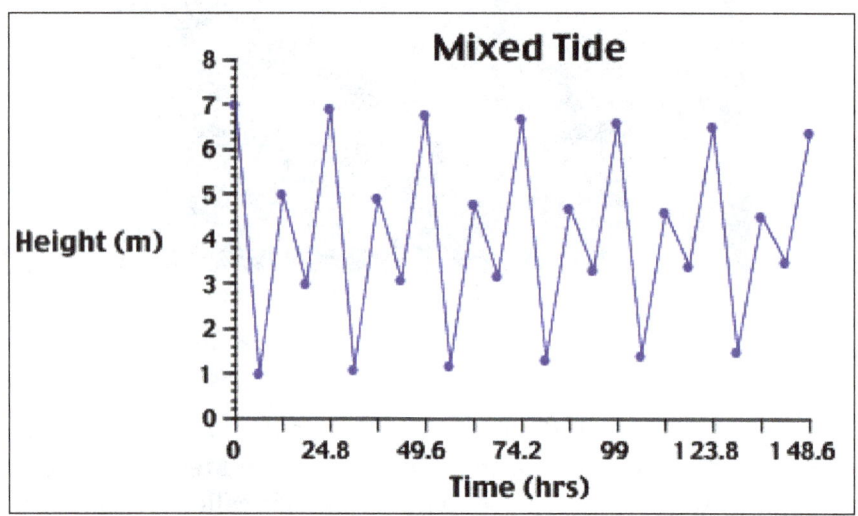

Figure: Cyclical tidal cycles associated with a mixed tide.

The map in below figure shows the geographic distribution of these three tide types on the Earth.

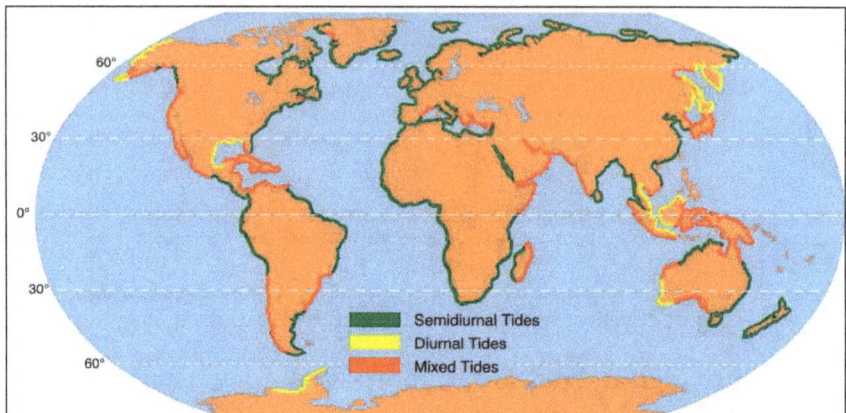

Figure: Global distribution of the three tidal types. Most of the world's coastlines have semidiurnal tides.

Chapter 5

Various Aspects of Ocean

The continuous and directed movement of ocean water produced by numerous factors that act upon water is known as ocean current. Some of the significant aspects associated with oceans are ocean acidification, ocean heat content, geophysical fluid dynamics, plate tectonics, ocean turbidity, etc. This chapter discusses in detail all these aspects of oceans.

Ocean Current

An ocean current is a continuous, directed movement of seawater generated by forces acting upon this mean flow, such as breaking waves, wind, the Coriolis effect, cabbeling, temperature and salinity differences, while tides are caused by the gravitational pull of the Sun and Moon. Depth contours, shoreline configurations, and interactions with other currents influence a current's direction and strength.

Ocean currents flow for great distances, and together, create the global conveyor belt which plays a dominant role in determining the climate of many of the Earth's regions. More specifically, ocean currents influence the temperature of the regions through which they travel. For example, warm currents traveling along more temperate coasts increase the temperature of the area by warming the sea breezes that blow over them. Perhaps the most striking example is the Gulf Stream, which makes northwest Europe much more temperate than any other region at the same latitude. Another example is Lima, Peru where the climate is cooler (sub-tropical) than the tropical latitudes in which the area is located, due to the effect of the Humboldt Current.

Function

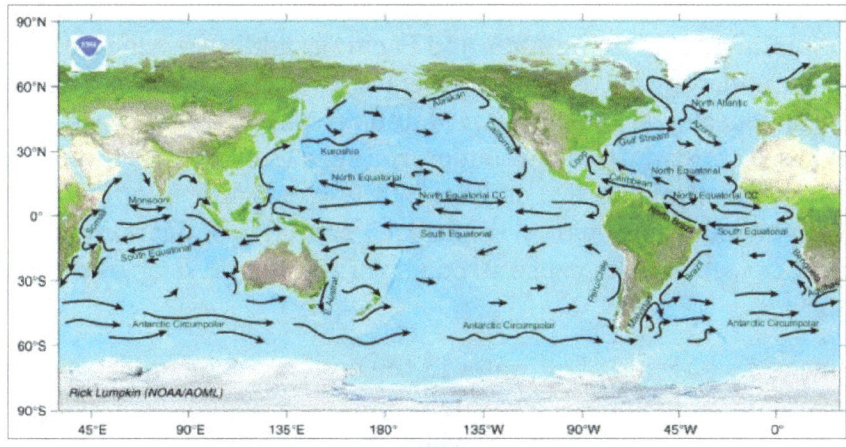

Major ocean surface currents (Source: NOAA).

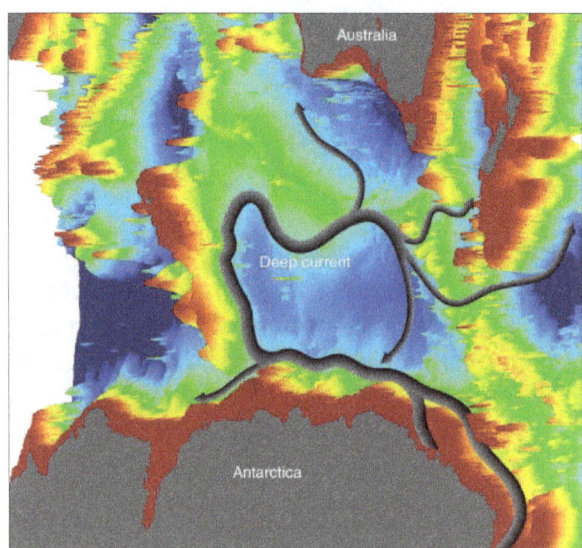

The bathymetry of the Kerguelen Plateau in the Southern Ocean governs the course of the new current part of the global network of ocean currents.

Surface oceanic currents are sometimes wind driven and develop their typical clockwise spirals in the northern hemisphere and counter-clockwise rotation in the southern hemisphere because of imposed wind stresses. In wind driven currents, the Ekman spiral effect results in the currents flowing at an angle to the driving winds. The areas of surface ocean currents move somewhat with the seasons; this is most notable in equatorial currents.

Deep ocean basins generally have a non-symmetric surface current, in that the eastern equatorward-flowing branch is broad and diffuse whereas the western poleward flowing branch is very narrow. These western boundary currents (of which the Gulf Stream is an example) are a consequence of the rotation of the Earth.

Deep ocean currents are driven by density and temperature gradients. Thermohaline circulation is also known as the ocean's conveyor belt (which refers to deep ocean density driven ocean basin currents). These currents, called submarine rivers, flow under the surface of the ocean and are hidden from immediate detection. Where significant vertical movement of ocean currents is observed, this is known as upwelling and downwelling. Deep ocean currents are currently being researched using a fleet of underwater robots called Argo.

The South Equatorial Currents of the Atlantic and Pacific straddle the equator. Though the Coriolis effect is weak near the equator (and absent at the equator), water moving in the currents on either side of the equator is deflected slightly poleward and replaced by deeper water. Thus, equatorial upwelling occurs in these westward flowing equatorial surface currents. Upwelling is an important process because this water from within and below the pycnocline is often rich in the nutrients needed by marine organisms for growth. By contrast, generally poor conditions for growth prevail in most of the open tropical ocean because strong layering isolates deep, nutrient rich water from the sunlit ocean surface.

Surface currents make up only 8% of all water in the ocean, are generally restricted to the upper 400 m (1,300 ft) of ocean water, and are separated from lower regions by varying temperatures and salinity which affect the density of the water, which in turn, defines each oceanic region. Because the movement of deep water in ocean basins is caused by density driven forces and gravity, deep waters sink into deep ocean basins at high latitudes where the temperatures are cold enough to cause the density to increase.

Ocean currents are measured in sverdrup (sv), where 1 sv is equivalent to a volume flow rate of 1,000,000 m³ (35,000,000 cu ft) per second.

Surface currents are found on the surface of an ocean, and are driven by large scale wind currents. They are directly affected by the wind—the Coriolis effect plays a role in their behaviors.

Thermohaline Circulation

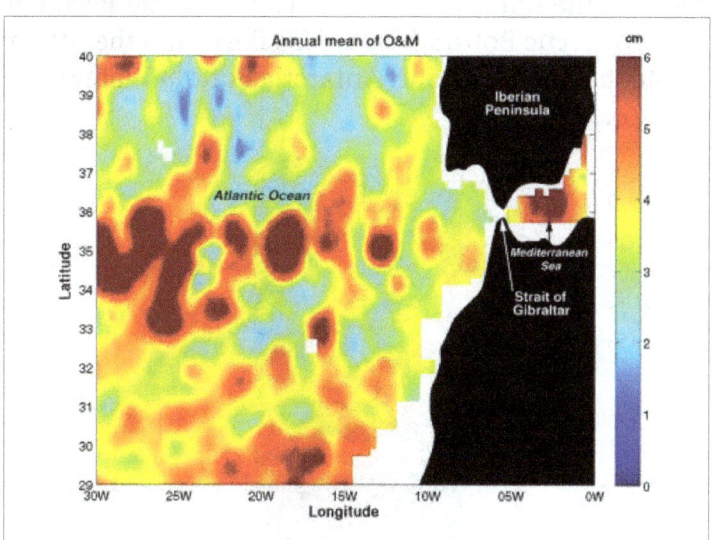

Coupling data collected by NASA/JPL by several different satellite-borne sensors, researchers have been able to "break through" the ocean's surface to detect "Meddies" -- super-salty warm-water eddies that originate in the Mediterranean Sea and then sink more than a half-mile underwater in the Atlantic Ocean. The Meddies are shown in red in this scientific figure.

Horizontal and vertical currents also exist below the pycnocline in the ocean's deeper waters. The movement of water due to differences in density as a function of water temperature and salinity is called thermohaline circulation. Ripple marks in sediments, scour lines, and the erosion of rocky outcrops on deep-ocean floors are evidence that relatively strong, localized bottom currents exist. Some of these currents may move as rapidly as 60 centimeters (24 inches) per second.

These currents are strongly influenced by bottom topography, since dense, bottom water must forcefully flow around seafloor projections. Thus, they are sometimes called contour currents. Bottom currents generally move equator-ward at or near the western boundaries of ocean basins (below the western boundary surface currents). The deep-water masses are not capable of moving water at speeds comparable to that of wind-driven surface currents. Water in some of these currents may move only 1 to 2 meters per day. Even at that slow speed, the Coriolis effect modifies their pattern of flow.

Downwelling of Deep Water in Polar Regions

Antarctic Bottom Water is the most distinctive of the deep-water masses. It is characterized by a salinity of 34.65‰, a temperature of -0.5 °C (30 °F), and a density of 1.0279 grams per cubic centimeter. This water is noted for its extreme density (the densest in the world ocean), for the great amount of it produced near Antarctic coasts, and for its ability to migrate north along the seafloor. Most Antarctic Bottom Water forms near the Antarctic coast south of South America during winter. Salt is concentrated in pockets between crystals of pure water and then squeezed out of the freezing mass to form a frigid brine. Between 20 million and 50 million cubic meters of this brine form every second. The water's great density causes it to sink toward the continental shelf, where it mixes with nearly equal parts of water from the southern Antarctic Circumpolar Current. The mixture settles along the edge of Antarctica's continental shelf, descends along the slope, and spreads along the deep-sea bed, creeping north in slow sheets. Antarctic Bottom Water flows many times as slowly as the water in surface currents: in the Pacific it may take a thousand years to reach the equator. Antarctic Bottom Water also flows into the Atlantic Ocean basin, where it flows north at a faster rate than in the Pacific. Antarctic Bottom Water has been identified as high as 40° N on the Atlantic floor.

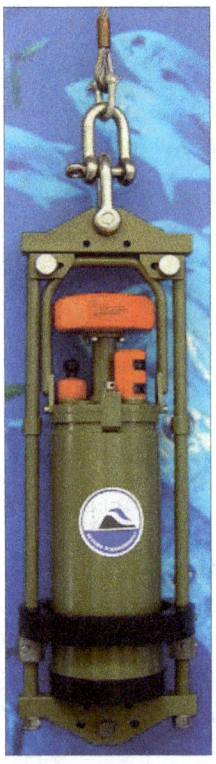

A recording current meter.

A small amount of dense bottom water also forms in the northern polar ocean. Although, the topography of the Arctic Ocean basin prevents most of the bottom water from escaping, with the exception of deep channels formed in the submarine ridges between Scotland, Iceland, and Greenland. These channels allow the cold, dense water formed in the Arctic to flow into the North Atlantic to form North Atlantic Deep Water. North Atlantic Deep Water forms when the relatively warm and salty North Atlantic Ocean cools as cold winds from northern Canada sweep over it. Exposed to the chilled air, water at the latitude of Iceland releases heat, cools from 10 °C to 2 °C, and sinks. Gulf Stream water that sinks in the north is replaced by warm water flowing clockwise along the U.S. east coast in the North Atlantic gyre.

Importance

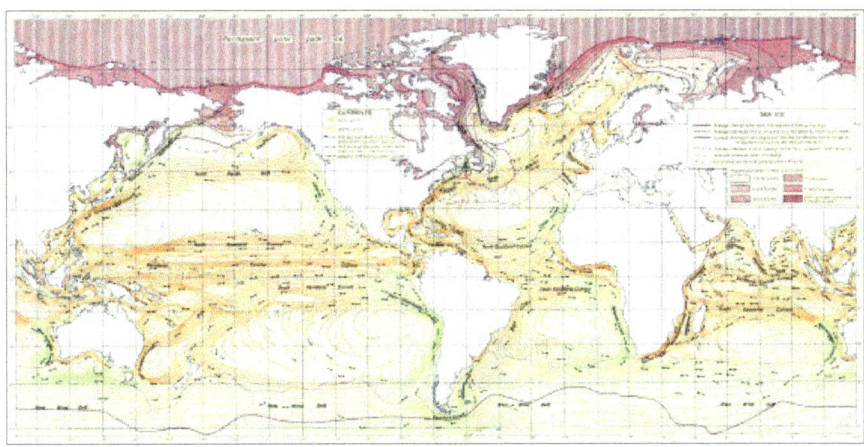

A 1943 map of the world's ocean currents.

Knowledge of surface ocean currents is essential in reducing costs of shipping, since traveling with them reduces fuel costs. In the wind powered sailing-ship era, knowledge was even more essential. A good example of this is the Agulhas Current, which long prevented Portuguese sailors from reaching India. In recent times, around-the-world sailing competitors make good use of surface currents to build and maintain speed. Ocean currents are also very important in the dispersal of many life forms. An example is the life-cycle of the European Eel.

Ocean currents are important in the study of marine debris, and vice versa. These currents also affect temperatures throughout the world. For example, the ocean current that brings warm water up the north Atlantic to northwest Europe also cumulatively and slowly blocks ice from forming along the seashores, which would also block ships from entering and exiting inland waterways and seaports, hence ocean currents play a decisive role in influencing the climates of regions through which they flow. Cold ocean water currents flowing from polar and sub-polar regions bring in a lot of plankton that are crucial to the continued survival of several key sea creature species in marine ecosystems. Since plankton are the food of fish, abundant fish populations often live where these currents prevail.

Ocean currents can also be used for marine power generation, with areas off of Japan, Florida and Hawaii being considered for test projects.

OSCAR: Near-realtime Global Ocean Surface Current Data Set

The OSCAR Near-realtime global ocean surface currents website from which users can create customized graphics and download the data. A section of the website provides validation studies in the form of graphics comparing OSCAR data with moored buoys and global drifters.

OSCAR data is used extensively in climate studies. maps and descriptions or annotations of climatic anomalies have been published in the monthly Climate Diagnostic Bulletin since 2001 and are routinely used to monitor ENSO and to test weather prediction models. OSCAR currents are routinely used to evaluate the surface currents in Global Circulation Models (GCMs), for example in NCEP Global Ocean Data Assimilation System (GODAS) and European Centre for Medium-Range Weather Forecasts (ECMWF).

Ocean Acidification

Ocean acidification is the ongoing decrease in the pH of the Earth's oceans, caused by the uptake of carbon dioxide (CO_2) from the atmosphere. Seawater is slightly basic (meaning pH > 7), and the process in question is a shift towards pH-neutral conditions rather than a transition to acidic conditions (pH < 7). Ocean alkalinity is not changed by the process, or may increase over long time periods due to carbonate dissolution. An estimated 30–40% of the carbon dioxide from human activity released into the atmosphere dissolves into oceans, rivers and lakes. To achieve chemical equilibrium, some of it reacts with the water to form carbonic acid. Some of these extra carbonic acid molecules react with a water molecule to give a bicarbonate ion and a hydronium ion, thus increasing ocean acidity (H^+ ion concentration). Between 1751 and 1994 surface ocean pH is estimated to have decreased from approximately 8.25 to 8.14, representing an increase of almost 30% in H^+ ion concentration in the world's oceans. Earth System Models project that within the last decade ocean acidity exceeded historical analogs and in combination with other ocean biogeochemical changes could undermine the functioning of marine ecosystems and disrupt the provision of many goods and services associated with the ocean.

Increasing acidity is thought to have a range of potentially harmful consequences for marine organisms, such as depressing metabolic rates and immune responses in some organisms, and causing coral bleaching. By increasing the presence of free hydrogen ions, each molecule of carbonic acid that forms in the oceans ultimately results in the conversion of *two* carbonate ions into bicarbonate ions. This net decrease in the amount of carbonate ions available makes it more difficult for marine calcifying organisms, such as coral and some plankton, to form biogenic calcium carbonate, and such structures become vulnerable to dissolution. Ongoing acidification of the oceans threatens food chains connected with the oceans. As members of the InterAcademy Panel, 105 science academies have issued a statement on ocean acidification recommending that by 2050, global CO_2 emissions be reduced by at least 50% compared to the 1990 level.

While ongoing ocean acidification is anthropogenic in origin, it has occurred previously in Earth's history. The most notable example is the Paleocene-Eocene Thermal Maximum (PETM), which occurred approximately 56 million years ago. For reasons that are currently uncertain, massive amounts of carbon entered the ocean and atmosphere, and led to the dissolution of carbonate sediments in all ocean basins.

Ocean acidification has been called the "evil twin of global warming" and "the other CO_2 problem".

Carbon Cycle

The carbon cycle describes the fluxes of carbon dioxide (CO 2) between the oceans, terrestrial biosphere, lithosphere, and the atmosphere. Human activities such as the combustion of fossil fuels and land use changes have led to a new flux of CO 2 into the atmosphere. About 45% has remained in the atmosphere; most of the rest has been taken up by the oceans, with some taken up by terrestrial plants.

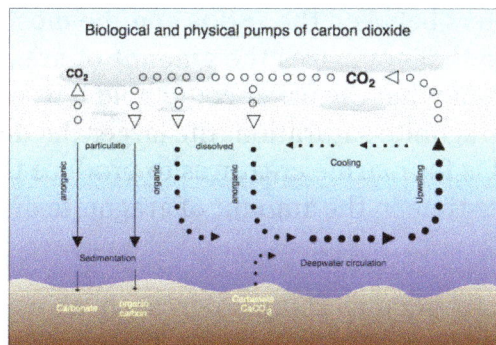

The CO 2 cycle between the atmosphere and the ocean.

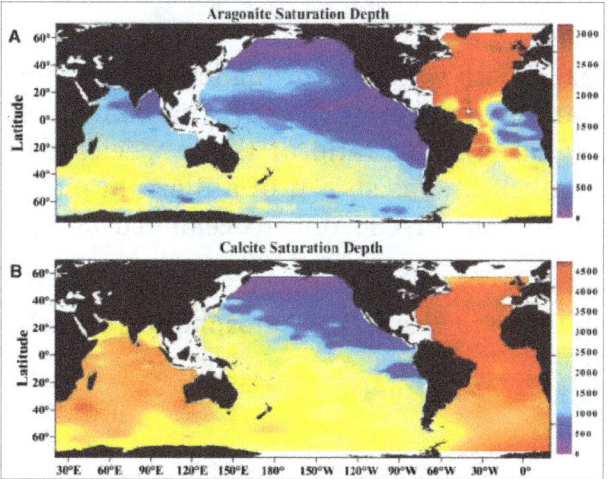

Distribution of (A) aragonite and (B) calcite saturation depth in the global oceans.

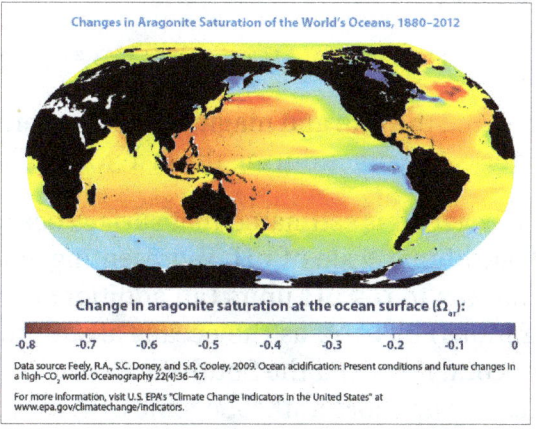

The map was created by the National Oceanic and Atmospheric Administration and the Woods Hole Oceanographic Institution using Community Earth System Model data. This map was created by comparing average conditions during the 1880s with average conditions during the most recent 10 years (2003–2012). Aragonite saturation has only been measured at selected locations during the last few decades, but it can be calculated reliably for different times and locations based on the relationships scientists have observed among aragonite saturation, pH, dissolved carbon, water temperature, concentrations of carbon dioxide in the atmosphere, and other factors that can be measured. This map shows changes in the amount of aragonite dissolved in ocean surface waters between the 1880s and the most recent decade (2003–2012). Aragonite saturation is a ratio that compares the amount of aragonite that is actually present with the total amount of aragonite that the water could hold if it were completely saturated. The more negative the change in aragonite saturation, the larger the decrease in aragonite available in the water, and the harder it is for marine creatures to produce their skeletons and shells. The global map shows changes over time in the amount of aragonite dissolved in ocean water, which is called aragonite saturation.

The carbon cycle involves both organic compounds such as cellulose and inorganic carbon compounds such as carbon dioxide and the carbonates. The inorganic compounds are particularly relevant when discussing ocean acidification for it includes many forms of dissolved CO_2 present in the Earth's oceans.

When CO_2 dissolves, it reacts with water to form a balance of ionic and non-ionic chemical species: dissolved free carbon dioxide ($CO_2(aq)$), carbonic acid (H_2CO_3), bicarbonate (HCO_3^-) and carbonate (CO_3^{2-}). The ratio of these species depends on factors such as seawater temperature and alkalinity (as shown in a Bjerrum plot). These different forms of dissolved inorganic carbon are transferred from an ocean's surface to its interior by the ocean's solubility pump.

The resistance of an area of ocean to absorbing atmospheric CO_2 is known as the Revelle factor.

Acidification

Dissolving CO_2 in seawater increases the hydrogen ion (H^+) concentration in the ocean, and thus decreases ocean pH, as follows:

$$CO_{2\,(aq)} + H_2O\ H_2CO_3\ HCO_3^- + H^+\ CO_3^{2-} + 2\ H^+.$$

Caldeira and Wickett (2003) placed the rate and magnitude of modern ocean acidification changes in the context of probable historical changes during the last 300 million years.

Since the industrial revolution began, it is estimated that surface ocean pH has dropped by slightly more than 0.1 units on the logarithmic scale of pH, representing about a 29% increase in H^+. It is expected to drop by a further 0.3 to 0.5 pH units (an additional doubling to tripling of today's post-industrial acid concentrations) by 2100 as the oceans absorb more anthropogenic CO_2, the impacts being most severe for coral reefs and the Southern Ocean. These changes are predicted to continue rapidly as the oceans take up more anthropogenic CO_2 from the atmosphere. The degree of change to ocean chemistry, including ocean pH, will depend on the mitigation and emissions pathways society takes.

Although the largest changes are expected in the future, a report from NOAA scientists found large quantities of water undersaturated in aragonite are already upwelling close to the Pacific continental shelf area of North America. Continental shelves play an important role in marine ecosystems since most marine organisms live or are spawned there, and though the study only dealt with the area from Vancouver to Northern California, the authors suggest that other shelf areas may be experiencing similar effects.

Average surface ocean pH				
Time	**pH**	**pH change relative to pre-industrial**	**Source**	**H⁺ concentration change relative to pre-industrial**
Pre-industrial (18th century)	8.179		analysed field	
Recent past (1990s)	8.104	−0.075	field	+ 18.9%
Present levels	~8.069	−0.11	field	+ 28.8%
2050 (2×CO2 = 560 ppm)	7.949	−0.230	model	+ 69.8%
2100 (IS92a)	7.824	−0.355	model	+ 126.5%

Rate

One of the first detailed datasets to examine how pH varied over a period of time at a temperate coastal location found that acidification was occurring much faster than previously predicted, with consequences for near-shore benthic ecosystems. Thomas Lovejoy, former chief biodiversity advisor to the World Bank, has suggested that "the acidity of the oceans will more than double in the next 40 years. This rate is 100 times faster than any changes in ocean acidity in the last 20 million years, making it unlikely that marine life can somehow adapt to the changes." It is predicted that, by the year 2100, the level of acidity in the ocean will reach the levels experienced by the earth 20 million years ago.

Current rates of ocean acidification have been compared with the greenhouse event at the Paleocene–Eocene boundary (about 55 million years ago) when surface ocean temperatures rose by 5–6 degrees Celsius. No catastrophe was seen in surface ecosystems, yet bottom-dwelling organisms in the deep ocean experienced a major extinction. The current acidification is on a path to reach levels higher than any seen in the last 65 million years, and the rate of increase is about ten times the rate that preceded the Paleocene–Eocene mass extinction. The current and projected acidification has been described as an almost unprecedented geological event. A National Research Council study released in April 2010 likewise concluded that "the level of acid in the oceans is increasing at an unprecedented rate." A 2012 paper in the journal *Science* examined the geological record in an attempt to find a historical analog for current global conditions as well as those of the future. The researchers determined that the current rate of ocean acidification is faster than at any time in the past 300 million years.

A review by climate scientists at the RealClimate blog, of a 2005 report by the Royal Society of the UK similarly highlighted the centrality of the *rates* of change in the present anthropogenic acidification process, writing:

"The natural pH of the ocean is determined by a need to balance the deposition and burial of $CaCO_3$ on the sea floor against the influx of $Ca2+$ and $CO2−3$ into the ocean from dissolving rocks on land, called weathering. These processes stabilize the pH of the ocean, by a mechanism called $CaCO_3$ compensation...The point of bringing it up again is to note that if the CO_2 concentration of the atmosphere changes more slowly than this, as it always has throughout the Vostok record, the pH of the ocean will be relatively unaffected because $CaCO_3$ compensation can keep up. The [present] fossil fuel acidification is much faster than natural changes, and so the acid spike will be more intense than the earth has seen in at least 800,000 years."

In the 15-year period 1995–2010 alone, acidity has increased 6 percent in the upper 100 meters of the Pacific Ocean from Hawaii to Alaska. According to a statement in July 2012 by Jane Lubchenco, head of the U.S. National Oceanic and Atmospheric Administration "surface waters are changing much more rapidly than initial calculations have suggested. It's yet another reason to be very seriously concerned about the amount of carbon dioxide that is in the atmosphere now and the additional amount we continue to put out."

A 2013 study claimed acidity was increasing at a rate 10 times faster than in any of the evolutionary crises in Earth's history. In a synthesis report published in *Science* in 2015, 22 leading marine scientists stated that CO_2 from burning fossil fuels is changing the oceans' chemistry more rapidly than at any time since the Great Dying, Earth's most severe known extinction event, emphasizing that the 2 °C maximum temperature increase agreed upon by governments reflects too small a cut in emissions to prevent "dramatic impacts" on the world's oceans, with lead author Jean-Pierre Gattuso remarking that "The ocean has been minimally considered at previous climate negotiations. Our study provides compelling arguments for a radical change at the UN conference (in Paris) on climate change".

Calcification

Overview

Changes in ocean chemistry can have extensive direct and indirect effects on organisms and their habitats. One of the most important repercussions of increasing ocean acidity relates to the production of shells and plates out of calcium carbonate (CaCO3). This process is called calcification and is important to the biology and survival of a wide range of marine organisms. Calcification involves the precipitation of dissolved ions into solid CaCO3 structures, such as coccoliths. After they are formed, such structures are vulnerable to dissolution unless the surrounding seawater contains saturating concentrations of carbonate ions (CO_3^{2-}).

Mechanism

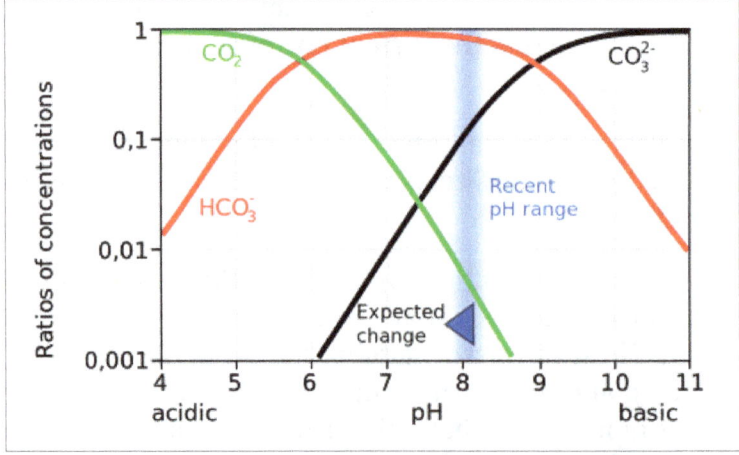

Bjerrum plot: Change in carbonate system of seawater from ocean acidification.

Of the extra carbon dioxide added into the oceans, some remains as dissolved carbon dioxide, while the rest contributes towards making additional bicarbonate (and additional carbonic acid). This also increases the concentration of hydrogen ions, and the percentage increase in hydrogen is larger than the percentage increase in bicarbonate, creating an imbalance in the reaction $HCO_3^- \rightarrow CO_3^{2-} + H^+$. To maintain chemical equilibrium, some of the carbonate ions already in the ocean combine with some of the hydrogen ions to make further bicarbonate. Thus the ocean's concentration of carbonate ions is reduced, creating an imbalance in the reaction $Ca^{2+} + CO_3^{2-} \rightarrow CaCO_3$, and making the dissolution of formed CaCO3 structures more likely.

These increases in concentrations of dissolved carbon dioxide and bicarbonate, and reduction in carbonate, are shown in a Bjerrum plot.

Saturation State

The saturation state (known as Ω) of seawater for a mineral is a measure of the thermodynamic potential for the mineral to form or to dissolve, and is described by the following equation:

$$\Omega = \frac{\left[Ca^{2+} \right]\left[CO_3^{2-} \right]}{K_{sp}}$$

Here Ω is the product of the concentrations (or activities) of the reacting ions that form the mineral (Ca^{2+} and CO_{2-3}), divided by the product of the concentrations of those ions when the mineral is at equilibrium (K_{sp}), that is, when the mineral is neither forming nor dissolving. In seawater, a natural horizontal boundary is formed as a result of temperature, pressure, and depth, and is known as the saturation horizon, or lysocline. Above this saturation horizon, Ω has a value greater than 1, and CaCO3 does not readily dissolve. Most calcifying organisms live in such waters. Below this depth, Ω has a value less than 1, and CaCO3 will dissolve. However, if its production rate is high enough to offset dissolution, CaCO3 can still occur where Ω is less than 1. The carbonate compensation depth occurs at the depth in the ocean where production is exceeded by dissolution.

The decrease in the concentration of CO_3^{2-} decreases Ω, and hence makes CaCO3 dissolution more likely.

Calcium carbonate occurs in two common polymorphs (crystalline forms): aragonite and calcite. Aragonite is much more soluble than calcite, so the aragonite saturation horizon is always nearer to the surface than the calcite saturation horizon. This also means that those organisms that produce aragonite may be more vulnerable to changes in ocean acidity than those that produce calcite. Increasing CO2 levels and the resulting lower pH of seawater decreases the saturation state of CaCO3 and raises the saturation horizons of both forms closer to the surface. This decrease in saturation state is believed to be one of the main factors leading to decreased calcification in marine organisms, as the inorganic precipitation of CaCO3 is directly proportional to its saturation state.

Possible Impacts

Increasing acidity has possibly harmful consequences, such as depressing metabolic rates in jumbo squid, depressing the immune responses of blue mussels, and coral bleaching. However it may benefit some species, for example increasing the growth rate of the sea star, *Pisaster ochraceus*, while shelled plankton species may flourish in altered oceans.

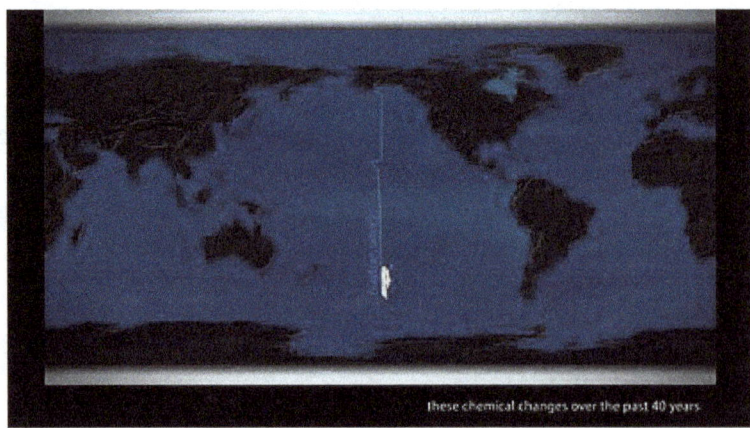

Video summarizing the impacts of ocean acidification. Source: NOAA Environmental Visualization Laboratory.

The report "Ocean Acidification Summary for Policymakers 2013" describes research findings and possible impacts.

Impacts on Oceanic Calcifying Organisms

Although the natural absorption of CO2 by the world's oceans helps mitigate the climatic effects of anthropogenic emissions of CO2, it is believed that the resulting decrease in pH will have negative consequences, primarily for oceanic calcifying organisms. These span the food chain from autotrophs to heterotrophs and include organisms such as coccolithophores, corals, foraminifera, echinoderms, crustaceans and molluscs. As described above, under normal conditions, calcite and aragonite are stable in surface waters since the carbonate ion is at supersaturating concentrations. However, as ocean pH falls, the concentration of carbonate ions required for saturation to occur increases, and when carbonate becomes undersaturated, structures made of calcium carbonate are vulnerable to dissolution. Therefore, even if there is no change in the rate of calcification, the rate of dissolution of calcareous material increases.

Corals, coccolithophore algae, coralline algae, foraminifera, shellfish and pteropods experience reduced calcification or enhanced dissolution when exposed to elevated CO2.

The Royal Society published a comprehensive overview of ocean acidification, and its potential consequences, in June 2005. However, some studies have found different response to ocean acidification, with coccolithophore calcification and photosynthesis both increasing under elevated atmospheric pCO_2, an equal decline in primary production and calcification in response to elevated CO_2 or the direction of the response varying between species. A study in 2008 examining a sediment core from the North Atlantic found that while the species composition of coccolithophorids has remained unchanged for the industrial period 1780 to 2004, the calcification of coccoliths has increased by up to 40% during the same time. A 2010 study from Stony Brook University suggested that while some areas are overharvested and other fishing grounds are being restored, because of ocean acidification it may be impossible to bring back many previous shellfish populations. While the full ecological consequences of these changes in calcification are still uncertain, it appears likely that many calcifying species will be adversely affected.

When exposed in experiments to pH reduced by 0.2 to 0.4, larvae of a temperate brittlestar, a relative of the common sea star, fewer than 0.1 percent survived more than eight days. There is also a suggestion that a decline in the coccolithophores may have secondary effects on climate, contributing to global warming by decreasing the Earth's albedo via their effects on oceanic cloud cover. All marine ecosystems on Earth will be exposed to changes in acidification and several other ocean biogeochemical changes.

The fluid in the internal compartments where corals grow their exoskeleton is also extremely important for calcification growth. When the saturation rate of aragonite in the external seawater is at ambient levels, the corals will grow their aragonite crystals rapidly in their internal compartments, hence their exoskeleton grows rapidly. If the level of aragonite in the external seawater is lower than the ambient level, the corals have to work harder to maintain the right balance in the internal compartment. When that happens, the process of growing the crystals slows down, and this slows down the rate of how much their exoskeleton is growing. Depending on how much aragonite is in the surrounding water, the corals may even stop growing because the levels of aragonite are too low to pump in to the internal compartment. They could even dissolve faster than they can make the crystals to their skeleton, depending on the aragonite levels in the surrounding water.

Ocean acidification may force some organisms to reallocate resources away from productive endpoints such as growth in order to maintain calcification.

In some places carbon dioxide bubbles out from the sea floor, locally changing the pH and other aspects of the chemistry of the seawater. Studies of these carbon dioxide seeps have documented a variety of responses by different organisms. Coral reef communities located near carbon dioxide seeps are of particular interest because of the sensitivity of some corals species to acidification. In Papua New Guinea, declining pH caused by carbon dioxide seeps is associated with declines in coral species diversity. However, in Palau carbon dioxide seeps are not associated with reduced species diversity of corals, although bioerosion of coral skeletons is much higher at low pH sites.

Other Biological Impacts

Aside from the slowing and/or reversing of calcification, organisms may suffer other adverse effects, either indirectly through negative impacts on food resources, or directly as reproductive or physiological effects. For example, the elevated oceanic levels of CO_2 may produce CO 2-induced acidification of body fluids, known as hypercapnia. Also, increasing ocean acidity is believed to have a range of direct consequences. For example, increasing acidity has been observed to: reduce metabolic rates in jumbo squid; depress the immune responses of blue mussels; and make it harder for juvenile clownfish to tell apart the smells of non-predators and predators, or hear the sounds of their predators. This is possibly because ocean acidification may alter the acoustic properties of seawater, allowing sound to propagate further, and increasing ocean noise. Calcium carbonate ions are very important when it comes to building organisms as they use calcium to build skeletons and shells. OA affects ocean species to a varying degree. Many plants do well with high CO_2 levels, but organisms that calcify like clams, mussels, corals and sea urchins might not do so well because everything in the ocean is connected including the food web. Since OA is such a large event because the ocean takes up 71% of the earth's surface OA is being referred to as climate change because it will affect the whole planet if the ocean becomes acidic. This impacts all animals that use sound for echolocation or communication. Atlantic longfin squid eggs took longer to hatch in acidified water, and the squid's statolith was smaller and malformed in animals placed in sea water with a lower pH. The lower PH was simulated with 20-30 times the normal amount of CO_2.

However, as with calcification, as yet there is not a full understanding of these processes in marine organisms or ecosystems.

Another possible effect would be an increase in red tide events, which could contribute to the accumulation of toxins (domoic acid, brevetoxin, saxitoxin) in small organisms such as anchovies and shellfish, in turn increasing occurrences of amnesic shellfish poisoning, neurotoxic shellfish poisoning and paralytic shellfish poisoning.

Nonbiological Impacts

Leaving aside direct biological effects, it is expected that ocean acidification in the future will lead to a significant decrease in the burial of carbonate sediments for several centuries, and even the dissolution of existing carbonate sediments. This will cause an elevation of ocean alkalinity, leading to the enhancement of the ocean as a reservoir for CO_2 with implications for climate change as more CO_2 leaves the atmosphere for the ocean.

Impact on Human Industry

The threat of acidification includes a decline in commercial fisheries and in the Arctic tourism industry and economy. Commercial fisheries are threatened because acidification harms calcifying organisms which form the base of the Arctic food webs.

Pteropods and brittle stars both form the base of the Arctic food webs and are both seriously damaged from acidification. Pteropods shells dissolve with increasing acidification and the brittle stars lose muscle mass when re-growing appendages. For pteropods to create shells they require aragonite which is produced through carbonate ions and dissolved calcium. Pteropods are severely affected because increasing acidification levels have steadily decreased the amount of water supersaturated with carbonate which is needed for aragonite creation. Arctic waters are changing so rapidly that they will become undersaturated with aragonite as early as 2016. Additionally the brittle star's eggs die within a few days when exposed to expected conditions resulting from Arctic acidification. Acidification threatens to destroy Arctic food webs from the base up. Arctic food webs are considered simple, meaning there are few steps in the food chain from small organisms to larger predators. For example, pteropods are "a key prey item of a number of higher predators - larger plankton, fish, seabirds, whales" Both pteropods and sea stars serve as a substantial food source and their removal from the simple food web would pose a serious threat to the whole ecosystem. The effects on the calcifying organisms at the base of the food webs could potentially destroy fisheries. The value of fish caught from US commercial fisheries in 2007 was valued at $3.8 billion and of that 73% was derived from calcifiers and their direct predators. Other organisms are directly harmed as a result of acidification. For example, decrease in the growth of marine calcifiers such as the American Lobster, Ocean Quahog, and scallops means there is less shellfish meat available for sale and consumption. Red king crab fisheries are also at a serious threat because crabs are calcifiers and rely on carbonate ions for shell development. Baby red king crab when exposed to increased acidification levels experienced 100% mortality after 95 days. In 2006 Red King Cab accounted for 23% of the total guideline harvest levels and a serious decline in red crab population would threaten the crab harvesting industry. Several ocean goods and services are likely to be undermined by future ocean acidification potentially affecting the livelihoods of some 400 to 800 million people depending upon the emission scenario.

Impact on Indigenous Peoples

Acidification could damage the Arctic tourism economy and affect the way of life of indigenous peoples. A major pillar of Arctic tourism is the sport fishing and hunting industry. The sport fishing industry is threatened by collapsing food webs which provide food for the prized fish. A decline in tourism lowers revenue input in the area, and threatens the economies that are increasingly dependent on tourism. Acidification is not merely a threat but has significantly declined whole fish populations. For example, In Scandinavia studies conducted on acidic water revealed that 15% of species populations had disappeared and that many more populations were limited in numbers or declining. The rapid decrease or disappearance of marine life could also affect the diet of Indigenous peoples.

Possible Responses

Reducing CO_2 Emissions

Members of the InterAcademy Panel recommended that by 2050, global anthropogenic CO_2 emissions be reduced less than 50% of the 1990 level. The 2009 statement also called on world leaders to:

- Acknowledge that ocean acidification is a direct and real consequence of increasing atmospheric CO_2 concentrations, is already having an effect at current concentrations, and is likely to cause grave harm to important marine ecosystems as CO_2 concentrations reach 450 [parts-per-million (ppm)] and above;

- [...] Recognise that reducing the build up of CO_2 in the atmosphere is the only practicable solution to mitigating ocean acidification;

- [...] Reinvigorate action to reduce stressors, such as overfishing and pollution, on marine ecosystems to increase resilience to ocean acidification.

Stabilizing atmospheric CO_2 concentrations at 450 ppm would require near-term emissions reductions, with steeper reductions over time.

The German Advisory Council on Global Change stated:

In order to prevent disruption of the calcification of marine organisms and the resultant risk of fundamentally altering marine food webs, the following guard rail should be obeyed: the pH of near surface waters should not drop more than 0.2 units below the pre-industrial average value in any larger ocean region (nor in the global mean).

One policy target related to ocean acidity is the magnitude of future global warming. Parties to the United Nations Framework Convention on Climate Change (UNFCCC) adopted a target of limiting warming to below 2 °C, relative to the pre-industrial level. Meeting this target would require substantial reductions in anthropogenic CO_2 emissions.

Limiting global warming to below 2 °C would imply a reduction in surface ocean pH of 0.16 from pre-industrial levels. This would represent a substantial decline in surface ocean pH.

Climate Engineering

Climate engineering (mitigating temperature or pH effects of emissions) has been proposed as a possible response to ocean acidification. The IAP (2009) statement cautioned against climate engineering as a policy response:

Mitigation approaches such as adding chemicals to counter the effects of acidification are likely to be expensive, only partly effective and only at a very local scale, and may pose additional unanticipated risks to the marine environment. There has been very little research on the feasibility and impacts of these approaches. Substantial research is needed before these techniques could be applied.

Reports by the WGBU (2006), the UK's Royal Society (2009), and the US National Research Council (2011) warned of the potential risks and difficulties associated with climate engineering.

Iron Fertilization

Iron fertilization of the ocean could stimulate photosynthesis in phytoplankton. The phytoplankton would convert the ocean's dissolved carbon dioxide into carbohydrate and oxygen gas, some of which would sink into the deeper ocean before oxidizing. More than a dozen open-sea experiments confirmed that adding iron to the ocean increases photosynthesis in phytoplankton by up to 30 times. While this approach has been proposed as a potential solution to the ocean acidification problem, mitigation of surface ocean acidification might increase acidification in the less-inhabited deep ocean.

A report by the UK's Royal Society (2009) reviewed the approach for effectiveness, affordability, timeliness and safety. The rating for affordability was "medium", or "not expected to be very cost-effective." For the other three criteria, the ratings ranged from "low" to "very low" (i.e., not good). For example, in regards to safety, the report found a "[high] potential for undesirable ecological side effects," and that ocean fertilization "may increase anoxic regions of ocean ('dead zones')."

Carbon Negative Fuels

Carbonic acid can be extracted from seawater as carbon dioxide for use in making synthetic fuel. If the resulting flue exhaust gas was subject to carbon capture, then the process would be carbon negative over time, resulting in permanent extraction of inorganic carbon from seawater and the atmosphere with which seawater is in equilibrium. Based on the energy requirements, this process was estimated to cost about $50 per tonne of CO_2.

Ocean Heat Content

Oceanic heat content (OHC) is the heat stored in the ocean. Oceanography and climatology are the science branches which study ocean heat content. Changes in the ocean heat content play an important role in the sea level rise, because of thermal expansion. It is with high confidence that ocean warming accounts for 90% of the energy accumulation from global warming between 1971 and 2010.

Definition and Measurement

The areal density of ocean heat content between two depth levels is defined as:

$$H = \rho c_p \int_{h2}^{h1} T(z)dz$$

Where ρ is seawater density, c_p is the specific heat of sea water, h2 is the lower depth, h1 is the upper depth, and $T(z)$ is the temperature profile. In SI units, H has units of $J \cdot m^{-2}$. Multiplying this quantity by the area of an ocean basin, or entire ocean, gives the total heat content, as indicated in the figure to right.

Ocean heat content can be computed using temperature measurements obtained by a Nansen bottle, an ARGO float, or ocean acoustic tomography. The World Ocean Database Project is the largest database for temperature profiles from all of the world's oceans.

Recent Changes

Several studies in recent years have found a multidecadal increase in OHC of the deep and upper ocean regions and attribute the heat uptake to anthropogenic warming. Studies based on *ARGO* indicate that ocean surface winds, especially the subtropical trade winds in the Pacific Ocean, change ocean heat vertical distribution. This results in changes among ocean currents, and an increase of the subtropical overturning, which is also related to the El Niño and La Niña phenomenon. Depending on stochastic natural variability fluctuations, during La Niña years around 30% more heat from the upper ocean layer is transported into the deeper ocean. Model studies indicate that ocean currents transport more heat into deeper layers during La Niña years, following changes in wind circulation. Years with increased ocean heat uptake have been associated with negative phases of the interdecadal Pacific oscillation (IPO). This is of particular interest to climate scientists who use the data to estimate the *ocean heat uptake*.

A study in 2015 concluded that ocean heat content increases by the Pacific Ocean, were compensated by an abrupt distribution of OHC into the Indian Ocean.

Role of Ocean Heat Content in Sea level Rise

Sea Level Rise

Sea level rise has been estimated to be on average between +2.6 millimetres (0.10 in) and 2.9 millimetres (0.11 in) per year ± 0.4 millimetres (0.016 in) since 1993.

Additionally, sea level rise has accelerated in recent years. For the period between 1870 and 2004, global average sea levels are estimated to have risen a total of 195 millimetres (7.7 in), and 1.7 millimetres (0.067 in) ± 0.3 millimetres (0.012 in) per year, with a significant acceleration of sea-level rise of 0.013 millimetres (0.00051 in) ± 0.006 millimetres (0.00024 in) per year per year.

According to one study of measurements available from 1950 to 2009, these measurements show an average annual rise in sea level of 1.7 millimetres (0.067 in) ± 0.3 millimetres (0.012 in) per year during this period, with satellite data showing a rise of 3.3 millimetres (0.13 in) ± 0.4 millimetres (0.016 in) per year from 1993 to 2009. Sea level rise is one of several lines of evidence that support the view that the global climate has recently warmed. In 2014 the USGCRP National

Climate Assessment projected that by the year 2100, the average sea level rise will have been between one and four feet (300mm-1200mm) since the date of the 2014 assessment. Current rates of sea level rise have roughly doubled since the pre 1992 rates of sea level rise of the 20th century.

In 2007, the Intergovernmental Panel on Climate Change (IPCC) stated that it is very likely human-induced (anthropogenic) warming contributed to the sea level rise observed in the latter half of the 20th century. The 2013 IPCC report (AR5) concluded, *"there is high confidence that the rate of sea level rise has increased during the last two centuries, and it is likely that GMSL (Global Mean Sea Level) has accelerated since the early 1900's.*

Sea level rises can considerably influence human populations in coastal and island regions and natural environments like marine ecosystems. Sea level rise is expected to continue for centuries. Because of the slow inertia, long response time for parts of the climate system, it has been estimated that we are already committed to a sea-level rise of approximately 2.3 metres (7.5 ft) for each degree Celsius of temperature rise within the next 2,000 years. It has been suggested that besides CO2 emissions reductions, a short term action to reduce sea level rise is to cut emissions of heat trapping gases such as methane and particulates such as soot.

Mechanism

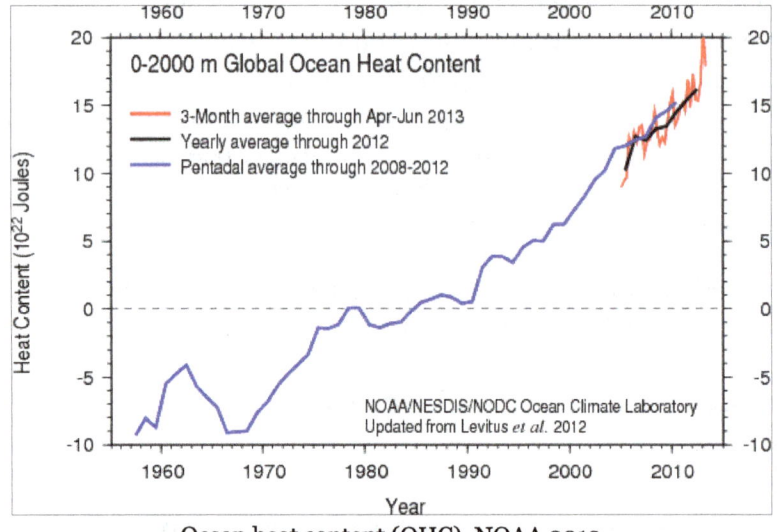

Ocean heat content (OHC), NOAA 2012.

There are two main mechanisms that contribute to observed sea level rise: (1) thermal expansion: because of the increase in ocean heat content (ocean water expands as it warms); and (2) the melting of major stores of land ice like ice sheets and glaciers.

On the timescale of centuries to millennia, the melting of ice sheets could result in even higher sea level rise. Partial deglaciation of the Greenland ice sheet, and possibly the West Antarctic ice sheet, could contribute 4 to 6 m (13 to 20 ft) or more to sea level rise.

Past Changes in Sea Level

Various factors affect the volume or mass of the ocean, leading to long-term changes in eustatic sea level. The two primary influences are temperature (because the density of water depends on temperature), and the mass of water locked up on land and sea as fresh water in rivers, lakes, glaciers and polar ice caps. Over much longer geological timescales, changes in the shape of oceanic basins and in land–sea distribution affect sea level. Since the Last Glacial Maximum about 20,000 years ago, sea level has risen by more than 125 m, with rates varying from tenths of a mm/yr to 10+mm/year, as a result of melting of major ice sheets.

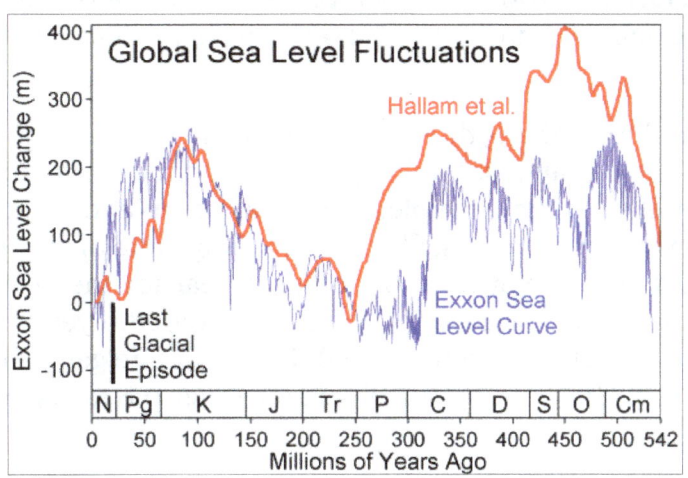

Comparison of two sea level reconstructions during the last 500 Ma. The scale of change during the last glacial/interglacial transition is indicated with a black bar. Note that over most of geologic history, long-term average sea level has been significantly higher than today.

During deglaciation between about 19,000 and 8,000 calendar years ago, sea level rose at extremely high rates as the result of the rapid melting of the British-Irish Sea, Fennoscandian, Laurentide, Barents-Kara, Patagonian, Innuitian ice sheets and parts of the Antarctic ice sheet. At the onset of deglaciation about 19,000 calendar years ago, a brief, at most 500-year long, glacio-eustatic event may have contributed as much as 10 m to sea level with an average rate of about 20 mm/yr. During the rest of the early Holocene, the rate of sea level rise varied from a low of about 6.0 - 9.9 mm/yr to as high as 30 - 60 mm/yr during brief periods of accelerated sea level rise.

Solid geological evidence, based largely upon analysis of deep cores of coral reefs, exists only for 3 major periods of accelerated sea level rise, called *meltwater pulses*, during the last deglaciation. They are Meltwater pulse 1A between circa 14,600 and 14,300 calendar years ago; Meltwater pulse 1B between circa 11,400 and 11,100 calendar years ago; and Meltwater pulse 1C between 8,200 and 7,600 calendar years ago. Meltwater pulse 1A was a 13.5 m rise over about 290 years centered at 14,200 calendar years ago and Meltwater pulse 1B was a 7.5 m rise over about 160 years centered at 11,000 years calendar years ago. In sharp contrast, the period between 14,300 and 11,100 calendar years ago, which includes the Younger Dryas interval, was an interval of reduced sea level rise at about 6.0 - 9.9 mm/yr. Meltwater pulse 1C was centered at 8,000 calendar years and produced a rise of 6.5 m in less than 140 year. Such rapid rates of sea level rising during meltwater events clearly implicate major ice-loss events related to ice sheet collapse. The primary source may have been meltwater from the Antarctic ice sheet. Other studies suggest a Northern Hemisphere source for the meltwater in the Laurentide ice sheet.

Recently, it has become widely accepted that late Holocene, 3,000 calendar years ago to present, sea level was nearly stable prior to an acceleration of rate of rise that is variously dated between

1850 and 1900 AD. Late Holocene rates of sea level rise have been estimated using evidence from archaeological sites and late Holocene tidal marsh sediments, combined with tide gauge and satellite records and geophysical modeling. For example, this research included studies of Roman wells in Caesarea and of Roman *piscinae* in Italy. These methods in combination suggest a mean eustatic component of 0.07 mm/yr for the last 2000 years.

Since 1880, as the Industrial Revolution took center stage, the ocean began to rise briskly, climbing a total of 210 mm (8.3 in) through 2009 causing extensive erosion worldwide and costing billions.

Sea level rose by 6 cm during the 19th century and 19 cm in the 20th century. Evidence for this includes geological observations, the longest instrumental records and the observed rate of 20th century sea level rise. For example, geological observations indicate that during the last 2,000 years, sea level change was small, with an average rate of only 0.0–0.2 mm per year. This compares to an average rate of 1.7 ± 0.5 mm per year for the 20th century. Baart et al. (2012) show that it is important to account for the effect of the 18.6-year lunar nodal cycle before acceleration in sea level rise should be concluded. Based on tide gauge data, the rate of global average sea level rise during the 20th century lies in the range 0.8 to 3.3 mm/yr, with an average rate of 1.8 mm/yr.

A two degrees Celsius of warming would warm the Earth above Eemian levels, move conditions closer to the Pliocene climate, a time when sea level was in the range of 25 meters higher than today. However, one study argues that sea level during the Pliocene might have only risen by 9 to 13.5 meters, due to more resilient ice sheets. Warren Cornwall, in: 'Ghosts of Ocean Past', published in an 'Science' monographic issue, 13 November 2015: 'Sea changes', pgs 752-755, presented a chart showing the current warming respect to preindustrial era of 1 °C, that goes along with the current CO_2 in atmosphere of 400 ppm. With the same 400 ppm CO_2, 3 million years ago, with an increased average temperature of 2 to 3 °C above our preindustrial levels, Sea level was between 6 meters and a not defined enough upper range. The issue may be not if Sea Level will raise, but how much, and at what a pace.

Projections

This graph shows the projected change in global sea level rise if atmospheric carbon dioxide (CO_2) concentrations were to either quadruple or double. The projection is based on several multi-century integrations of a GFDL global coupled ocean-atmosphere model. These projections are the expected changes due to thermal expansion of sea water alone, and do not include the effect of melted continental ice sheets. With the effect of ice sheets included, the total rise could be larger by a substantial factor. Image credit: NOAA GFDL.

21st Century

The 2007 Fourth Assessment Report (IPCC 4) projected century-end sea levels using the Special Report on Emissions Scenarios (SRES). SRES developed emissions scenarios to project climate-change impacts. The projections based on these scenarios are not predictions, but reflect plausible estimates of future social and economic development (e.g., economic growth, population level). The six SRES "marker" scenarios projected sea level to rise by 18 to 59 centimetres (7.1 to 23.2 in).

Their projections were for the time period 2090–99, with the increase in level relative to average sea level over the 1980–99 period. This estimate did not include all of the possible contributions of ice sheets.

Hansen (2007), assumed an ice sheet contribution of 1 cm for the decade 2005–15, with a potential ten year doubling time for sea-level rise, based on a nonlinear ice sheet response, which would yield 5 m this century.

Research from 2008 observed rapid declines in ice-mass balance from both Greenland and Antarctica, and concluded that sea-level rise by 2100 is likely to be at least twice as large as that presented by IPCC AR4, with an upper limit of about two meters.

Projections assessed by the US National Research Council (2010) suggest possible sea level rise over the 21st century of between 56 and 200 cm (22 and 79 in). The NRC describes the IPCC projections as "conservative".

In 2011, Rignot and others projected a rise of 32 centimetres (13 in) by 2050. Their projection included increased contributions from the Antarctic and Greenland ice sheets. Use of two completely different approaches reinforced the Rignot projection.

In its Fifth Assessment Report (2013), The IPCC found that recent observations of global average sea level rise at a rate of 3.2 [2.8 to 3.6] mm per year is consistent with the sum of contributions from observed thermal ocean expansion due to rising temperatures (1.1 [0.8 to 1.4] mm per year), glacier melt (0.76 [0.39 to 1.13] mm per year), Greenland ice sheet melt (0.33 [0.25 to 0.41] mm per year), Antarctic ice sheet melt (0.27 [0.16 to 0.38] mm per year), and changes to land water storage (0.38 [0.26 to 0.49] mm per year). The report had also concluded that if emissions continue to keep up with the worst case IPCC scenarios, global average sea level could rise by nearly 1m by 2100 (0.52–0.98 m from a 1986-2005 baseline). If emissions follow the lowest emissions scenario, then global average sea level is projected to rise by between 0.28–0.6 m by 2100 (compared to a 1986–2005 baseline).

The Third National Climate Assessment (NCA), released May 6, 2014, projected a sea level rise of 1 to 4 feet by 2100 (30–120 cm). Decision makers who are particularly susceptible to risk may wish to use a wider range of scenarios from 8 inches to 6.6 feet by 2100.

A 2015 study by sea level rise experts concluded that based on MIS 5e data, sea level rise could rise faster in the coming decades, with a doubling time of 10, 20 or 40 years. The study abstract explains: *We argue that ice sheets in contact with the ocean are vulnerable to non-linear disintegration in response to ocean warming, and we posit that ice sheet mass loss can be approximated by a doubling time up to sea level rise of at least several meters. Doubling times of 10, 20 or 40 years yield sea level rise of several meters in 50, 100 or 200 years. Paleoclimate data reveal that subsurface ocean warming causes ice shelf melt and ice sheet discharge.*

Our climate model exposes amplifying feedbacks in the Southern Ocean that slow Antarctic bottom water formation and increase ocean temperature near ice shelf grounding lines, while cooling the surface ocean and increasing sea ice cover and water column stability. Ocean surface cooling, in the North Atlantic as well as the Southern Ocean, increases tropospheric horizontal temperature gradients, eddy kinetic energy and baroclinicity, which drive more powerful storms. However, Greg Holland from the National Center for Atmospheric Research, who reviewed the study, noted *"There is no doubt that the sea level rise, within the IPCC, is a very conservative number, so the truth lies somewhere between IPCC and Jim."*

After 2100

There is a widespread consensus that substantial long-term sea-level rise will continue for centuries to come. IPCC AR4 estimated that at least a partial deglaciation of the Greenland ice sheet, and possibly the West Antarctic ice sheet, would occur given a global average temperature increase of 1–4 °C (relative to temperatures over the years 1990–2000). This estimate was given about a 50% chance of being correct. The estimated timescale was centuries to millennia, and would contribute 4 to 6 metres (13 to 20 ft) or more to sea levels over this period.

Models

There is the possibility of a rapid change in glaciers, ice sheets, and hence sea level. Predictions of such a change are highly uncertain due to a lack of scientific understanding. Modeling of the processes associated with a rapid ice-sheet and glacier change could potentially increase future projections of sea-level rise.

Hansen (2007), concluded that paleoclimate ice sheet models generally do not include physics of ice streams, effects of surface melt descending through crevasses and lubricating basal flow, or realistic interactions with the ocean. The calibration of projected modelling for future sea-level rise is generally done with a linear projection of future sea level. Thus, does not include potential nonlinear collapse of an ice sheet.

Contribution

Close-up of Ross Ice Shelf, the largest ice shelf of Antarctica, about the size of
France and up to several hundred metres thick.

Each year about 8 mm of precipitation (liquid equivalent) falls on the ice sheets in Antarctica and Greenland, mostly as snow, which accumulates and over time forms glacial ice. Much of this precipitation began as water vapor evaporated from the ocean surface. To a first approximation, the same amount of water appeared to return to the ocean in icebergs and from ice melting at the edges. Scientists previously had estimated which is greater, ice going in or coming out, called the mass balance, important because a nonzero balance causes changes in global sea level. High-precision gravimetry from satellites determined that Greenland was losing more than 200 billion tons of ice per year, in accord with loss estimates from ground measurement. The rate of ice loss was accelerating, having grown from 137 gigatons in 2002–2003.

- The total global ice mass lost from Greenland, Antarctica and Earth's glaciers and ice caps during 2003–2010 was about 4.3 trillion tons (1,000 cubic miles), adding about 12 mm (0.5 in) to global sea level, enough ice to cover an area comparable to the United States 50 cm (1.5 ft) deep.

- The melting of small glaciers on the margins of Greenland and the Antarctic Peninsula would increase sea level around 0.5 meter. At the extreme potential, according to the Third Assessment Report of the International Panel on Climate Change, the ice contained within the Greenland ice sheet entirely melted increases sea level by 7.2 meters (24 feet). The ice contained within the Antarctic ice sheet entirely melted would produce 61.1 meters (200 feet) of sea-level change, both totaling a sea-level rise of 68.3 meters (224 feet).

It is estimated that Antarctica, if fully melted, would contribute more than 60 metres of sea level rise, and Greenland would contribute more than 7 metres. Small glaciers and ice caps on the margins of Greenland and the Antarctic Peninsula might contribute about 0.5 metres. While the latter figure is much smaller than for Antarctica or Greenland it could occur relatively quickly (within the coming century) whereas melting of Greenland would be slow (perhaps 1,500 years to fully deglaciate at the fastest likely rate) and Antarctica even slower. However, this calculation does not account for the possibility that as meltwater flows under and lubricates the larger ice sheets, they could begin to move much more rapidly towards the sea.

In 2002, Rignot and Thomas found that the West Antarctic and Greenland ice sheets were losing mass, while the East Antarctic ice sheet was probably in balance (although they could not determine the sign of the mass balance for The East Antarctic ice sheet). Kwok and Comiso (*J. Climate*, v15, 487–501, 2002) also discovered that temperature and pressure anomalies around West Antarctica and on the other side of the Antarctic Peninsula correlate with recent Southern Oscillation events.

In 2005 it was reported that during 1992–2003, East Antarctica thickened at an average rate of about 18 mm/yr while West Antarctica showed an overall thinning of 9 mm/yr. associated with increased precipitation. A gain of this magnitude is enough to slow sea-level rise by 0.12 ± 0.02 mm/yr.

Antarctica

On the Antarctic continent itself, the large volume of ice present stores around 70% of the world's fresh water. This ice sheet is constantly gaining ice from snowfall and losing ice through outflow to the sea.

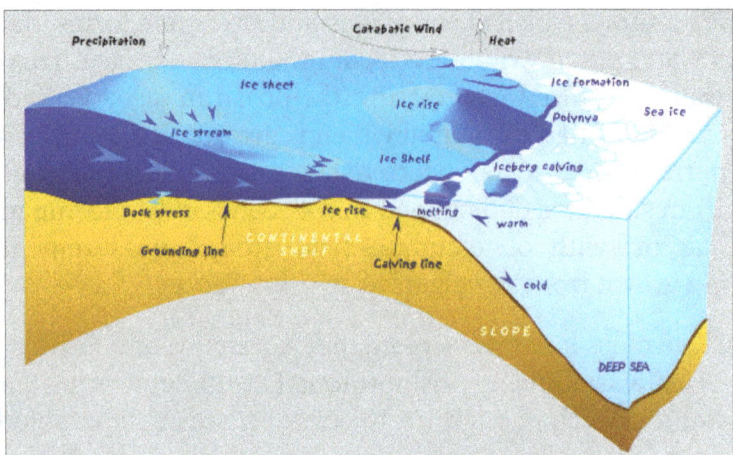

Processes around an Antarctic ice shelf.

Sheperd et al. 2012, found that different satellite methods were in good agreement and combing methods leads to more certainty with East Antarctica, West Antarctica, and the Antarctic Peninsula changing in mass by +14 ± 43, −65 ± 26, and −20 ± 14 gigatonnes per year.

East Antarctic Ice Sheet (EAIS)

East Antarctica is a cold region with a ground-base above sea level and occupies most of the continent. This area is dominated by small accumulations of snowfall which becomes ice and thus eventually seaward glacial flows. The mass balance of the East Antarctic Ice Sheet as a whole over the period 1980-2004 is thought to be slightly positive (lowering sea level) or near to balance, with a large degree of uncertainty. However, increased ice outflow has been suggested in some regions.

West Antarctic Ice Sheet (WAIS)

West Antarctica is currently experiencing a net outflow of glacial ice, which will increase global sea level over time. A review of the scientific studies looking at data from 1992 to 2006 suggested a net loss of around 50 gigatons of ice per year was a reasonable estimate (around 0.14 mm of sea-level rise), although significant acceleration of outflow glaciers in the Amundsen Sea Embayment could have more than doubled this figure for the year 2006.

Thomas et al. found evidence of an accelerated contribution to sea level rise from West Antarctica. The data showed that the Amundsen Sea sector of the West Antarctic Ice Sheet was discharging 250 cubic kilometres of ice every year, which was 60% more than precipitation accumulation in the catchment areas. This alone was sufficient to raise sea level at 0.24 mm/yr. Further, thinning rates for the glaciers studied in 2002–03 had increased over the values measured in the early 1990s. The bedrock underlying the glaciers was found to be hundreds of metres deeper than previously known, indicating exit routes for ice from further inland in the Byrd Subpolar Basin. Thus the West Antarctic ice sheet may not be as stable as has been supposed.

A 2009 study found that the rapid collapse of West Antarctic Ice Sheet would raise sea level by 3.3 metres (11 ft).

Glaciers

Observational and modelling studies of mass loss from glaciers and ice caps indicate a contribution to sea-level rise of 0.2–0.4 mm/yr, averaged over the 20th century. The results from Dyurgerov show a sharp increase in the contribution of mountain and subpolar glaciers to sea-level rise since 1996 (0.5 mm/yr) to 1998 (2 mm/yr) with an average of about 0.35 mm/yr since 1960. Of interest also is Arendt et al., who estimate the contribution of Alaskan glaciers of 0.14±0.04 mm/yr between the mid-1950s to the mid-1990s, increasing to 0.27 mm/yr in the middle and late 1990s.

Greenland

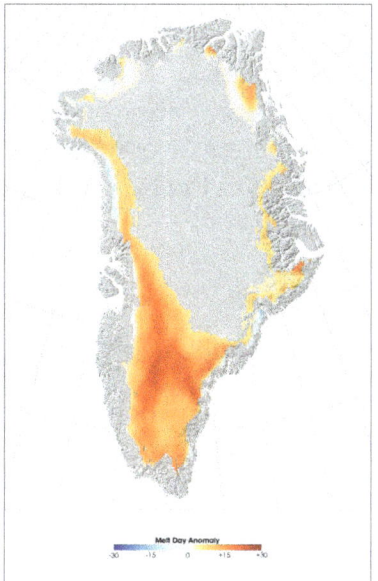

Greenland 2007 melt anomaly, measured as the difference between the number of days on which melting occurred in 2007 compared to the average annual melting days from 1988–2006.

In 2004 Rignot et al. estimated a contribution of 0.04 ± 0.01 mm/yr to sea level rise from South East Greenland. In the same year, Krabill *et al.* estimate a net contribution from Greenland to be at least 0.13 mm/yr in the 1990s. Joughin *et al.* have measured a doubling of the speed of Jakobshavn Isbræ between 1997 and 2003. This is Greenland's largest outlet glacier; it drains 6.5% of the ice sheet, and is thought to be responsible for increasing the rate of sea-level rise by about 0.06 millimetres per year, or roughly 4% of the 20th-century rate of sea-level increase. In 2004, Rignot *et al.* estimated a contribution of 0.04±0.01 mm/yr to sea-level rise from southeast Greenland.

Rignot and Kanagaratnam produced a comprehensive study and map of the outlet glaciers and basins of Greenland. They found widespread glacial acceleration below 66 N in 1996 which spread to 70 N by 2005; and that the ice sheet loss rate in that decade increased from 90 to 200 cubic km/yr; this corresponds to an extra 0.25–0.55 mm/yr of sea level rise.

In July 2005 it was reported that the Kangerlussuaq Glacier, on Greenland's east coast, was moving towards the sea three times faster than a decade earlier. Kangerdlugssuaq is around 1,000 m thick, 7.2 km (4.5 miles) wide, and drains about 4% of the ice from the Greenland ice sheet. Measurements of Kangerdlugssuaq in 1988 and 1996 showed it moving at between 5 and 6 km/yr (3.1–3.7 miles/yr), while in 2005 that speed had increased to 14 km/yr (8.7 miles/yr).

According to the 2004 Arctic Climate Impact Assessment, climate models project that local warming in Greenland will exceed 3 °C during this century. Also, ice-sheet models project that such a warming would initiate the long-term melting of the ice sheet, leading to a complete melting of the Greenland ice sheet over several millennia, resulting in a global sea level rise of about seven metres.

Subsidence and Effective Sea Level Rise

Many ports, urban conglomerations, and agricultural regions are built on river deltas, where subsidence contributes to a substantial increase in *effective* sea level rise. This is caused by both unsustainable extraction of groundwater (in some place also by extraction of oil and gas), and by levees and other flood management practices that prevent accumulation of sediments to compensate for the natural settling of deltaic soils. In many deltas this results in subsidence ranging from several millimeters per year up to possibly 25 centimeters per year in parts of the Ciliwung delta (Jakarta). Total anthropogenic-caused subsidence in the Rhine-Meuse-Scheldt delta (Netherlands) is estimated at 3 to 4 meters, over nine meters in the Sacramento-San Joaquin River Delta, and over ten feet in urban areas of the Mississippi River Delta (New Orleans).

Effects

Map of major cities of the world most vulnerable to sea level rise.

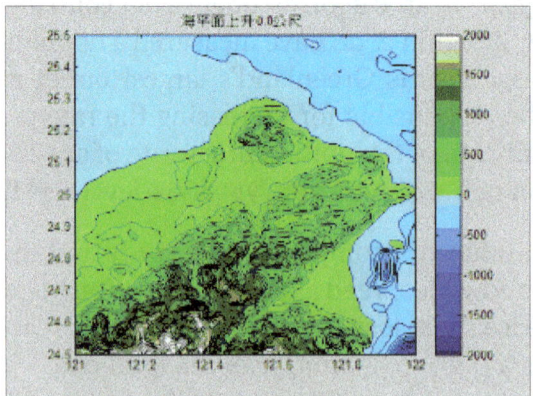

Schematic animation of sea level rise in Taipei, Taiwan and surrounding regions, in meters.

The IPCC TAR WGII report (*Impacts, Adaptation Vulnerability*) notes that current and future climate change would be expected to have a number of impacts, particularly on coastal systems. Such impacts may include increased coastal erosion, higher storm-surge flooding, inhibition of primary production processes, more extensive coastal inundation, changes in surface water quality and groundwater characteristics, increased loss of property and coastal habitats, increased flood risk and potential loss of life, loss of non-monetary cultural resources and values, impacts on agriculture and aquaculture through decline in soil and water quality, and loss of tourism, recreation, and transportation functions.

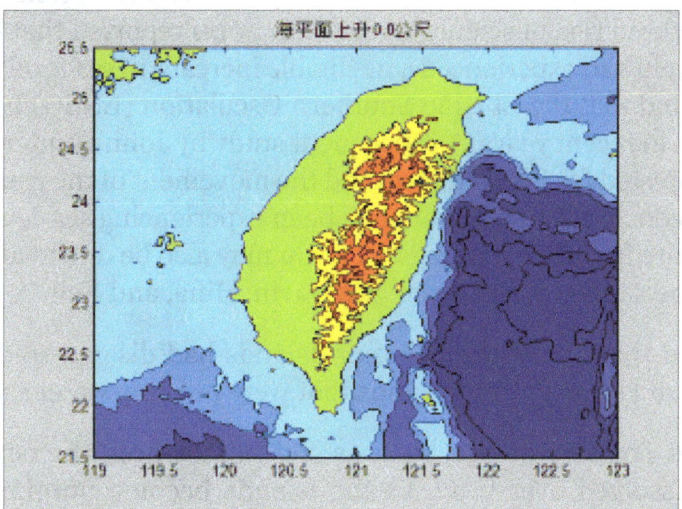

Schematic animation of sea level rise in Taiwan and surrounding regions, in meters.

There is an implication that many of these impacts will be detrimental—especially for the three-quarters of the world's poor who depend on agriculture systems. The report does, however, note that owing to the great diversity of coastal environments; regional and local differences in projected relative sea level and climate changes; and differences in the resilience and adaptive capacity of ecosystems, sectors, and countries, the impacts will be highly variable in time and space.

The IPCC report of 2007 estimated that accelerated melting of the Himalayan ice caps and the resulting rise in sea levels would likely increase the severity of flooding in the short term during the rainy season and greatly magnify the impact of tidal storm surges during the cyclone season. A sea-level rise of just 400 mm in the Bay of Bengal would put 11 percent of the Bangladesh's coastal land underwater, creating 7–10 million climate refugees.

Sea level rise could also displace many shore-based populations: for example it is estimated that a sea level rise of just 200 mm could make 740,000 people in Nigeria homeless.

Future sea-level rise, like the recent rise, is not expected to be globally uniform. Some regions show a sea-level rise substantially more than the global average (in many cases of more than twice the average), and others a sea level fall. However, models disagree as to the likely pattern of sea level change.

Island Nations

IPCC assessments suggest that deltas and small island states are particularly vulnerable to sea-level rise caused by both thermal expansion and increased ocean water. Sea level changes have not yet been conclusively proven to have directly resulted in environmental, humanitarian, or economic losses to small island states, but the IPCC and other bodies have found this a serious risk scenario in coming decades.

Maldives, Tuvalu, and other low-lying countries are among the areas that are at the highest level of risk. The UN's environmental panel has warned that, at current rates, sea level would be high enough to make the Maldives uninhabitable by 2100.

Many media reports have focused on the island nations of the Pacific, notably the Polynesian islands of Tuvalu, which based on more severe flooding events in recent years, were thought to be "sinking" due to sea level rise. A scientific review in 2000 reported that based on University of Hawaii gauge data, Tuvalu had experienced a negligible increase in sea level of 0.07 mm a year over the past two decades, and that the El Niño Southern Oscillation (ENSO) had been a larger factor in Tuvalu's higher tides in recent years. A subsequent study by John Hunter from the University of Tasmania, however, adjusted for ENSO effects and the movement of the gauge (which was thought to be sinking). Hunter concluded that Tuvalu had been experiencing sea-level rise of about 1.2 mm per year. The recent more frequent flooding in Tuvalu may also be due to an erosional loss of land during and following the actions of 1997 cyclones Gavin, Hina, and Keli.

In 2016 it was reported that five of the Solomon Islands had disappeared due to the combined effects of sea level rise and stronger trade winds that were pushing water into the Western Pacific.

Besides the issues that flooding brings, such as soil salinisation, the island states themselves would also become dissolved over time, as the islands become uninhabitable or completely submerged by the sea. Once this happens, all rights on the surrounding area (sea) are removed. This area can be huge as rights extend to a radius of 224 nautical miles (414 km) around the entire island state. Any resources, such as fossil oil, minerals and metals, within this area can be freely dug up by anyone and sold without needing to pay any commission to the (now dissolved) island state.

Options that have been proposed to assist island nations to adapt to rising sea level include abandoning islands, building dikes, and building upwards.

Cities

A study in the April, 2007 issue of *Environment and Urbanization* reports that 634 million people live in coastal areas within 30 feet (9.1 m) of sea level. The study also reported that about two thirds of the world's cities with over five million people are located in these low-lying coastal areas. Future sea level rise could lead to potentially catastrophic difficulties for shore-based communities in the next centuries: for example, many major cities such as Venice, London, New Orleans, and New York already need storm-surge defenses, and would need more if the sea level rose, though they also face issues such as subsidence. However, modest increases in sea level are likely to be offset when cities adapt by constructing sea walls or through relocating.

Re-insurance company Swiss Re estimates an economic loss for southeast Florida in 2030, of $33 billion from climate-related damages. Miami has been listed as "the number-one most vulnerable city worldwide" in terms of potential damage to property from storm-related flooding and sea-level rise.

Habitats

Coastal and Polar habitats are facing drastic changes as consequence of rising sea levels. Loss of ice in the Arctic may force local species to migrate in search of a new home. If seawater continues to approach inland, problems related to contaminated soils and flooded wetlands may occur. Also, fish, birds, and coastal plants could lose parts of their habitat. In 2016 it was reported that the Bramble Cay melomys, which lived on a Great Barrier Reef island, had probably become extinct because of sea level rises.

Extreme Sea Level Rise Events

Downturn of Atlantic meridional overturning circulation (AMOC), has been tied to extreme regional sea level rise (1-in-850 year event). Between 2009–2010, coastal sea levels north of New York City increased by 128 mm within two years. This jump is unprecedented in the tide gauge records, which collects data since a couple of centuries.

Sea Level Measurement

Satellites

Jason-1 continues the same sea surface measurements begun by TOPEX/Poseidon. It will be followed by the Ocean Surface Topography Mission on Jason-2 and by a planned future Jason-3.

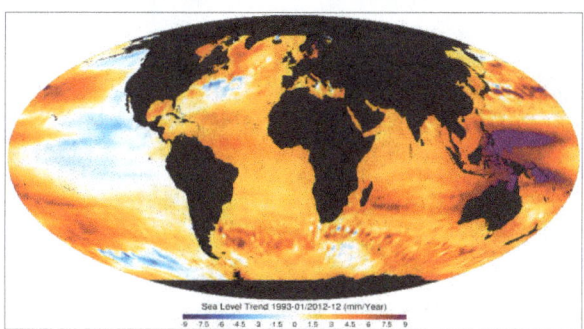

1993–2012 Sea level trends from satellite altimetry.

In 1992 the TOPEX/Poseidon satellite was launched to record the change in sea level. Current rates of sea level rise from satellite altimetry have been estimated in the range of 2.9–3.4 ± 0.4–0.6 mm per year for 1993–2010. This exceeds those from tide gauges. It is unclear whether this represents an increase over the last decades; variability; true differences between satellites and tide gauges; or problems with satellite calibration. Due to calibration errors of the first satellite – Topex/Poseidon, sea levels have been slightly overestimated until 2015, which resulted in masking of ongoing sea level rise acceleration.

Tide Gauge

Amsterdam

The longest running sea-level measurements, NAP or Amsterdam Ordnance Datum established in 1675, are recorded in Amsterdam, the Netherlands. About 25 percent of the Netherlands lies beneath sea level, while more than 50 percent of this nation's area would be inundated by temporary floods if it did not have an extensive levee system.

Australia

In Australia, data collected by the Commonwealth Scientific and Industrial Research Organisation (CSIRO) show the current global mean sea level trend to be 3.2 mm/yr., a doubling of the rate of the total increase of about 210mm that was measured from 1880 to 2009, which reflected an average annual rise over the entire 129-year period of about 1.6 mm/year.

Australian record collection has a long time horizon, including measurements by an amateur meteorologist beginning in 1837 and measurements taken from a sea-level benchmark struck on a small cliff on the Isle of the Dead near the Port Arthur convict settlement on 1 July 1841. These records, when compared with data recorded by modern tide gauges, reinforce the recent comparisons of the historic sea level rise of about 1.6 mm/year, with the sharp acceleration in recent decades.

Continuing extensive sea level data collection by Australia's (CSIRO) is summarized in in its finding of mean sea level trend to be 3.2 mm/yr. As of 2003 the National Tidal Centre of the Bureau of Meteorology managed 32 tide gauges covering the entire Australian coastline, with some measurements available starting in 1880.

United States

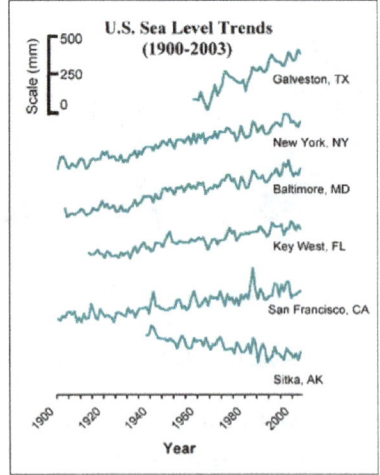

US sea-level trends 1900–2003.

Tide gauges in the United States reveal considerable variation because some land areas are rising and some are sinking. For example, over the past 100 years, the rate of sea level rise varied from about an increase of 0.36 inches (9.1 mm) per year along the Louisiana Coast (due to land sinking), to a drop of a few inches per decade in parts of Alaska (due to post-glacial rebound). The rate of sea level rise increased during the 1993–2003 period compared with the longer-term average (1961–2003), although it is unclear whether the faster rate reflected a short-term variation or an increase in the long-term trend.

One study showed no acceleration in sea level rise in US tide gauge records during the 20th century. However, another study found that the rate of rise for the US Atlantic coast during the 20th century was far higher than during the previous two thousand years.

Adaptation

In 2008, the Dutch *Delta Commission* (Deltacommissie), advised in a report that the Netherlands would need a massive new building program to strengthen the country's water defenses against the anticipated effects of global warming for the next 190 years. This commission was created in September 2007, after the damage caused by Hurricane Katrina prompted reflection and preparations. Those included drawing up worst-case plans for evacuations. The plan included more than €100 billion (US$144 bn), in new spending through the year 2100 to take measures, such as broadening coastal dunes and strengthening sea and river dikes. The commission said the country must plan for a rise in the North Sea up to 1.3 metres (4 ft 3 in) by 2100, rather than the previously projected 0.80 metres (2 ft 7 in), and plan for a 2–4 metre (6.5–13 feet) rise by 2200.

The New York City Panel on Climate Change (NPCC), is an effort to prepare the New York City area for climate change.

Miami Beach is spending $500 million in the next years to address sea-level rise. Actions include a pump drainage system, and to raise roadways and sidewalks.

Geophysical Fluid Dynamics

Geophysical fluid dynamics is the study of naturally occurring, large-scale flows on Earth and other planets. It is applied to the motion of fluids in the ocean and outer core, and to gases in the atmosphere of Earth and other planets. Two features that are common to many of the phenomena studied in geophysical fluid dynamics are rotation of the fluid due to the planetary rotation and stratification (layering). The applications of geophysical fluid dynamics do not generally include the circulation of the mantle, which is the subject of geodynamics, or fluid phenomena in the magnetosphere. Smaller scale flow features (those negligibly influenced by the rotation of the Earth) are the province of fields such as hydrology, physical oceanography and meteorology.

Fundamentals

To describe the flow of geophysical fluids, equations are needed for conservation of momentum (or Newton's second law) and conservation of energy. The former leads to the Navier-Stokes equations. Further approximations are generally made. First, the fluid is assumed to be incompressible. Remarkably, this works well even for a highly compressible fluid like air as long as sound and shock waves can be ignored. Second, the fluid is assumed to be a Newtonian fluid, meaning that there is a linear relation between the shear stress τ and the strain u, for example,

$$\tau = \mu \frac{du}{dx},$$

where μ is the viscosity. Under these assumptions the Navier-Stokes equations are,

$$\rho \Big(\underbrace{\frac{\partial \mathbf{v}}{\partial t}}_{\substack{\text{Eulerian} \\ \text{acceleration}}} + \underbrace{\mathbf{v} \cdot \nabla \mathbf{v}}_{\text{Advection}} \Big) = \underbrace{-\nabla p}_{\substack{\text{Pressure} \\ \text{gradient}}} + \underbrace{\mu \nabla^2 \mathbf{v}}_{\text{Viscosity}} + \underbrace{\mathbf{f}}_{\substack{\text{Other} \\ \text{body} \\ \text{forces}}} .$$

The left hand side represents the acceleration that a small parcel of fluid would experience in a reference frame that moved with the parcel (a Lagrangian frame of reference). In a stationary (Eulerian) frame of reference, this acceleration is divided into the local rate of change of velocity and advection, a measure of the rate of flow in or out of a small region.

The equation for energy conservation is essentially an equation for heat flow. If heat is transported by conduction, the heat flow is governed by a diffusion equation. If there are also buoyancy effects, for example hot air rising, then natural convection can occur.

Buoyancy and Stratification

Fluid that is less dense than its surroundings tends to rise until it has the same density as its surroundings. If there is not much energy input to the system, it will tend to become stratified. On a large scale, Earth's atmosphere is divided into a series of layers. Going upwards from the ground, these are the troposphere, stratosphere, mesosphere, thermosphere, and exosphere.

The density of air is mainly determined by temperature and water vapor content, the density of sea water by temperature and salinity, and the density of lake water by temperature. Where stratification occurs, there may be thin layers in which temperature or some other property changes more rapidly with height or depth than the surrounding fluid. Depending on the main sources of buoyancy, this layer may be called a pycnocline (density), thermocline (temperature), halocline (salinity), or chemocline (chemistry, including oxygenation).

The same buoyancy that gives rise to stratification also drives gravity waves. If the gravity waves occur within the fluid, they are called internal waves.

In modeling buoyancy-driven flows, the Navier-Stokes equations are modified using the Boussinesq approximation. This ignores variations in density except where they are multiplied by the gravitational acceleration g.

If the pressure depends only on density and vice versa, the fluid dynamics are called barotropic. In the atmosphere, this corresponds to a lack of fronts, as in the tropics. If there are fronts, the flow is baroclinic, and instabilities such as cyclones can occur.

Plate Tectonics

Plate tectonics is a scientific theory describing the large-scale motion of Earth's lithosphere. The theoretical model builds on the concept of continental drift developed during the first few decades of the 20th century. The geoscientific community accepted plate-tectonic theory after seafloor spreading was validated in the late 1950s and early 1960s.

The lithosphere, which is the rigid outermost shell of a planet (the crust and upper mantle), is broken up into tectonic plates. The Earth's lithosphere is composed of seven or eight major plates (depending on how they are defined) and many minor plates. Where the plates meet, their relative motion determines the type of boundary: convergent, divergent, or transform. Earthquakes, volcanic activity, mountain-building, and oceanic trench formation occur along these plate boundaries. The relative movement of the plates typically ranges from zero to 100 mm annually.

Tectonic plates are composed of oceanic lithosphere and thicker continental lithosphere, each topped by its own kind of crust. Along convergent boundaries, subduction carries plates into the mantle; the material lost is roughly balanced by the formation of new (oceanic) crust along divergent margins by seafloor spreading. In this way, the total surface of the lithosphere remains the same. This prediction of plate tectonics is also referred to as the conveyor belt principle. Earlier theories (that still have some supporters) propose gradual shrinking (contraction) or gradual expansion of the globe.

Tectonic plates are able to move because the Earth's lithosphere has greater strength than the underlying asthenosphere. Lateral density variations in the mantle result in convection. Plate movement is thought to be driven by a combination of the motion of the seafloor away from the spreading ridge (due to variations in topography and density of the crust, which result in differences in gravitational forces) and drag, with downward suction, at the subduction zones. Another explanation lies in the different forces generated by tidal forces of the Sun and Moon. The relative importance of each of these factors and their relationship to each other is unclear, and still the subject of much debate.

Key Principles

The outer layers of the Earth are divided into the lithosphere and asthenosphere. This is based on differences in mechanical properties and in the method for the transfer of heat. Mechanically, the lithosphere is cooler and more rigid, while the asthenosphere is hotter and flows more easily. In terms of heat transfer, the lithosphere loses heat by conduction, whereas the asthenosphere also transfers heat by convection and has a nearly adiabatic temperature gradient. This division should not be confused with the *chemical* subdivision of these same layers into the mantle (comprising both the asthenosphere and the mantle portion of the lithosphere) and the crust: a given piece of mantle may be part of the lithosphere or the asthenosphere at different times depending on its temperature and pressure.

The key principle of plate tectonics is that the lithosphere exists as separate and distinct *tectonic plates*, which ride on the fluid-like (visco-elastic solid) asthenosphere. Plate motions range up to a typical 10–40 mm/year (Mid-Atlantic Ridge; about as fast as fingernails grow), to about 160 mm/year (Nazca Plate; about as fast as hair grows). The driving mechanism behind this movement is described below.

Tectonic lithosphere plates consist of lithospheric mantle overlain by either or both of two types of crustal material: oceanic crust (in older texts called *sima* from silicon and magnesium) and continental crust (*sial* from silicon and aluminium). Average oceanic lithosphere is typically 100 km (62 mi) thick; its thickness is a function of its age: as time passes, it conductively cools and subjacent cooling mantle is added to its base. Because it is formed at mid-ocean ridges and spreads outwards, its thickness is therefore a function of its distance from the mid-ocean ridge where it was formed. For a typical distance that oceanic lithosphere must travel before being subducted, the thickness varies from about 6 km (4 mi) thick at mid-ocean ridges to greater than 100 km (62 mi) at subduction zones; for shorter or longer distances, the subduction zone (and therefore also the mean) thickness becomes smaller or larger, respectively. Continental lithosphere is typically ~200 km thick, though this varies considerably between basins, mountain ranges, and stable cratonic interiors of continents. The two types of crust also differ in thickness, with continental crust being considerably thicker than oceanic (35 km vs. 6 km).

The location where two plates meet is called a *plate boundary*. Plate boundaries are commonly associated with geological events such as earthquakes and the creation of topographic features such as mountains, volcanoes, mid-ocean ridges, and oceanic trenches. The majority of the world's active volcanoes occur along plate boundaries, with the Pacific Plate's Ring of Fire being the most active and widely known today. These boundaries are discussed in further detail below. Some volcanoes occur in the interiors of plates, and these have been variously attributed to internal plate deformation and to mantle plumes.

As explained above, tectonic plates may include continental crust or oceanic crust, and most plates contain both. For example, the African Plate includes the continent and parts of the floor of the Atlantic and Indian Oceans. The distinction between oceanic crust and continental crust is based on their modes of formation. Oceanic crust is formed at sea-floor spreading centers, and continental crust is formed through arc volcanism and accretion of terranes through tectonic processes, though some of these terranes may contain ophiolite sequences, which are pieces of oceanic crust considered to be part of the continent when they exit the standard cycle of formation and spreading centers and subduction beneath continents. Oceanic crust is also denser than continental crust owing to their different compositions. Oceanic crust is denser because it has less silicon and more heavier elements ("mafic") than continental crust ("felsic"). As a result of this density stratification, oceanic crust generally lies below sea level (for example most of the Pacific Plate), while continen-tal crust buoyantly projects above sea level.

Types of Plate Boundaries

Three types of plate boundaries exist, with a fourth, mixed type, characterized by the way the plates move relative to each other. They are associated with different types of surface phenomena. The different types of plate boundaries are:

1. *Transform boundaries (Conservative)* occur where two lithospheric plates slide, or perhaps more accurately, grind past each other along transform faults, where plates are neither created nor destroyed. The relative motion of the two plates is either sinistral (left side toward the observer) or dextral (right side toward the observer). Transform faults occur across a spreading center. Strong earthquakes can occur along a fault. The San Andreas Fault in California is an example of a transform boundary exhibiting dextral motion.

2. *Divergent boundaries (Constructive)* occur where two plates slide apart from each other. At zones of ocean-to-ocean rifting, divergent boundaries form by seafloor spreading, allowing for the formation of new ocean basin. As the continent splits, the ridge forms at the spreading center, the ocean basin expands, and finally, the plate area increases causing many small volcanoes and/or shallow earthquakes. At zones of continent-to-continent rifting, divergent boundaries may cause new ocean basin to form as the continent splits, spreads, the central rift collapses, and ocean fills the basin. Active zones of Mid-ocean ridges (e.g., Mid-Atlantic Ridge and East Pacific Rise), and continent-to-continent rifting (such as Africa's East African Rift and Valley, Red Sea) are examples of divergent boundaries.

3. *Convergent boundaries (Destructive)* (or *active margins*) occur where two plates slide toward each other to form either a subduction zone (one plate moving underneath the other) or a continental collision. At zones of ocean-to-continent subduction (e.g. the Andes mountain range in South America, and the Cascade Mountains in Western United States), the dense oceanic lithosphere plunges beneath the less dense continent. Earthquakes trace the path of the downward-moving plate as it descends into asthenosphere, a trench forms, and as the subducted plate is heated it releases volatiles, mostly water from hydrous minerals, into the surrounding mantle. The addition of water lowers the melting point of the mantle material above the subducting slab, causing it to melt. The magma that results typically leads to volcanism. At zones of ocean-to-ocean subduction (e.g. Aleutian islands, Mariana Islands, and the Japanese island arc), older, cooler, denser crust slips beneath less dense crust. This causes earthquakes and a deep trench to form in an arc shape. The upper mantle of the subducted plate then heats and magma rises to form curving chains of volcanic islands. Deep marine trenches are typically associated with subduction zones, and the basins that develop along the active boundary are often called "foreland basins". Closure of ocean basins can occur at continent-to-continent boundaries (e.g., Himalayas and Alps): collision between masses of granitic continental lithosphere; neither mass is subducted; plate edges are compressed, folded, uplifted.

4. *Plate boundary zones* occur where the effects of the interactions are unclear, and the boundaries, usually occurring along a broad belt, are not well defined and may show various types of movements in different episodes.

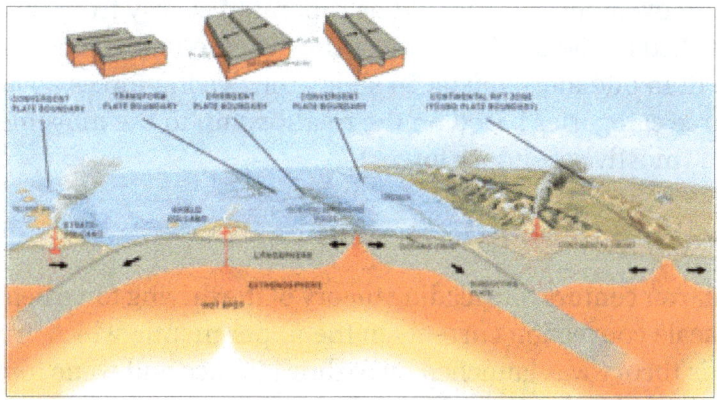

Three types of plate boundary.

Driving Forces of Plate Motion

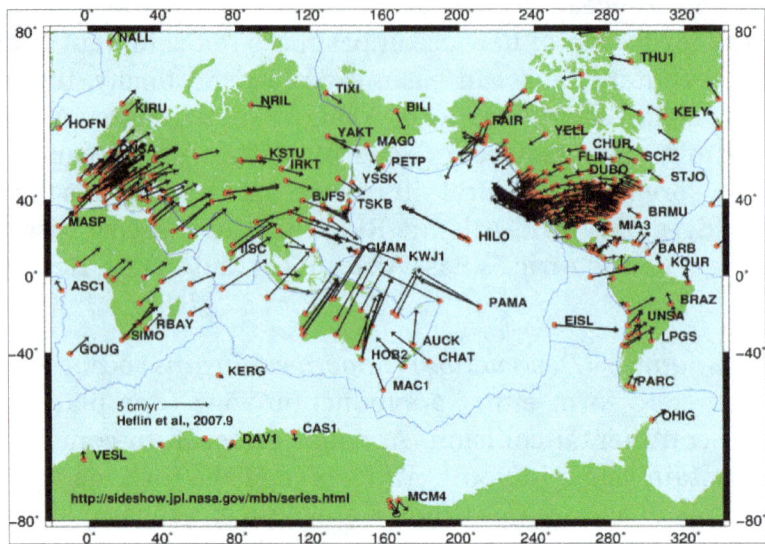

Plate motion based on Global Positioning System (GPS) satellite data from NASA JPL. The vectors show direction and magnitude of motion.

It is generally accepted that tectonic plates are able to move because of the relative density of oceanic lithosphere and the relative weakness of the asthenosphere. Dissipation of heat from the mantle is acknowledged to be the original source of the energy required to drive plate tectonics through convection or large scale upwelling and doming. The current view, though still a matter of some debate, asserts that as a consequence, a powerful source of plate motion is generated due to the excess density of the oceanic lithosphere sinking in subduction zones. When the new crust forms at mid-ocean ridges, this oceanic lithosphere is initially less dense than the underlying asthenosphere, but it becomes denser with age as it conductively cools and thickens. The greater density of old lithosphere relative to the underlying asthenosphere allows it to sink into the deep mantle at subduction zones, providing most of the driving force for plate movement. The weakness of the asthenosphere allows the tectonic plates to move easily towards a subduction zone. Although subduction is thought to be the strongest force driving plate motions, it cannot be the only force since there are plates such as the North American Plate which are moving, yet are nowhere being subducted. The same is true for the enormous Eurasian Plate. The sources of plate motion are a matter of intensive research and discussion among scientists. One of the main points is that the kinematic pattern of the movement itself should be separated clearly from the possible geodynamic mechanism that is invoked as the driving force of the observed movement, as some patterns may be explained by more than one mechanism. In short, the driving forces advocated at the moment can be divided into three categories based on the relationship to the movement: mantle dynamics related, gravity related (mostly secondary forces).

Driving Forces Related to Mantle Dynamics

For much of the last quarter century, the leading theory of the driving force behind tectonic plate motions envisaged large scale convection currents in the upper mantle which are transmitted through the asthenosphere. This theory was launched by Arthur Holmes and some forerunners in the 1930s and was immediately recognized as the solution for the acceptance of the theory as originally discussed in the papers of Alfred Wegener in the early years of the century. However, despite its acceptance, it was long debated in the scientific community because the leading ("fixist") theory still envisaged a static Earth without moving continents up until the major breakthroughs of the early sixties.

Two- and three-dimensional imaging of Earth's interior (seismic tomography) shows a varying lateral density distribution throughout the mantle. Such density variations can be material (from rock chemistry), mineral (from variations in mineral structures), or thermal (through thermal expansion and contraction from heat energy). The manifestation of this varying lateral density is mantle convection from buoyancy forces.

How mantle convection directly and indirectly relates to plate motion is a matter of ongoing study and discussion in geodynamics. Somehow, this energy must be transferred to the lithosphere for tectonic plates to move. There are essentially two types of forces that are thought to influence plate motion: friction and gravity.

- Basal drag (friction): Plate motion driven by friction between the convection currents in the asthenosphere and the more rigid overlying lithosphere.

- Slab suction (gravity): Plate motion driven by local convection currents that exert a downward pull on plates in subduction zones at ocean trenches. Slab suction may occur in a geodynamic setting where basal tractions continue to act on the plate as it dives into the mantle (although perhaps to a greater extent acting on both the under and upper side of the slab).

Lately, the convection theory has been much debated as modern techniques based on 3D seismic tomography still fail to recognize these predicted large scale convection cells. Therefore, alternative views have been proposed:

In the theory of plume tectonics developed during the 1990s, a modified concept of mantle convection currents is used. It asserts that super plumes rise from the deeper mantle and are the drivers or substitutes of the major convection cells. These ideas, which find their roots in the early 1930s with the so-called "fixistic" ideas of the European and Russian Earth Science Schools, find resonance in the modern theories which envisage hot spots/mantle plumes which remain fixed and are overridden by oceanic and continental lithosphere plates over time and leave their traces in the geological record (though these phenomena are not invoked as real driving mechanisms, but rather as modulators). Modern theories that continue building on the older mantle doming concepts and see plate movements as a secondary phenomena are beyond the scope of this page and are discussed elsewhere (for example on the plume tectonics page).

Another theory is that the mantle flows neither in cells nor large plumes but rather as a series of channels just below the Earth's crust, which then provide basal friction to the lithosphere. This theory, called "surge tectonics", became quite popular in geophysics and geodynamics during the 1980s and 1990s.

Driving Forces Related to Gravity

Forces related to gravity are usually invoked as secondary phenomena within the framework of a more general driving mechanism such as the various forms of mantle dynamics described above.

Gravitational sliding away from a spreading ridge: According to many authors, plate motion is driven by the higher elevation of plates at ocean ridges. As oceanic lithosphere is formed at spreading ridges from hot mantle material, it gradually cools and thickens with age (and thus adds distance from the ridge). Cool oceanic lithosphere is significantly denser than the hot mantle material from which it is derived and so with increasing thickness it gradually subsides into the mantle to compensate the greater load. The result is a slight lateral incline with increased distance from the ridge axis.

This force is regarded as a secondary force and is often referred to as "ridge push". This is a misnomer as nothing is "pushing" horizontally and tensional features are dominant along ridges. It is more accurate to refer to this mechanism as gravitational sliding as variable topography across the totality of the plate can vary considerably and the topography of spreading ridges is only the most prominent feature. Other mechanisms generating this gravitational secondary force include flexural bulging of the lithosphere before it dives underneath an adjacent plate which produces a clear topographical feature that can offset, or at least affect, the influence of topographical ocean ridges, and mantle plumes and hot spots, which are postulated to impinge on the underside of tectonic plates.

Slab-pull: Current scientific opinion is that the asthenosphere is insufficiently competent or rigid to directly cause motion by friction along the base of the lithosphere. Slab pull is therefore most widely thought to be the greatest force acting on the plates. In this current understanding, plate motion is mostly driven by the weight of cold, dense plates sinking into the mantle at trenches. Recent models indicate that trench suction plays an important role as well. However, as the North American Plate is nowhere being subducted, yet it is in motion presents a problem. The same holds for the African, Eurasian, and Antarctic plates.

Gravitational sliding away from mantle doming: According to older theories, one of the driving mechanisms of the plates is the existence of large scale asthenosphere/mantle domes which cause the gravitational sliding of lithosphere plates away from them. This gravitational sliding represents a secondary phenomenon of this basically vertically oriented mechanism. This can act on various scales, from the small scale of one island arc up to the larger scale of an entire ocean basin.

Driving Forces Related to Earth Rotation

Alfred Wegener, being a meteorologist, had proposed tidal forces and pole flight force as the main driving mechanisms behind continental drift; however, these forces were considered far too small to cause continental motion as the concept then was of continents plowing through oceanic crust. Therefore, Wegener later changed his position and asserted that convection currents are the main driving force of plate tectonics in the last edition of his book in 1929.

However, in the plate tectonics context (accepted since the seafloor spreading proposals of Heezen, Hess, Dietz, Morley, Vine, and Matthews during the early 1960s), oceanic crust is suggested to be in motion *with* the continents which caused the proposals related to Earth rotation to be reconsidered. In more recent literature, these driving forces are:

1. Tidal drag due to the gravitational force the Moon (and the Sun) exerts on the crust of the Earth;

2. Global deformation of the geoid due to small displacements of rotational pole with respect to the Earth's crust;

3. Other smaller deformation effects of the crust due to wobbles and spin movements of the Earth rotation on a smaller time scale.

Forces that are small and generally negligible are:

1. The Coriolis force.

2. The centrifugal force, which is treated as a slight modification of gravity.

For these mechanisms to be overall valid, systematic relationships should exist all over the globe between the orientation and kinematics of deformation and the geographical latitudinal and longitudinal grid of the Earth itself. Ironically, these systematic relations studies in the second half of the nineteenth century and the first half of the twentieth century underline exactly the opposite: that the plates had not moved in time, that the deformation grid was fixed with respect to the Earth equator and axis, and that gravitational driving forces were generally acting vertically and caused only local horizontal movements (the so-called pre-plate tectonic, "fixist theories"). Later studies (discussed below on this page), therefore, invoked many of the relationships recognized during this pre-plate tectonics period to support their theories.

Of the many forces discussed in this paragraph, tidal force is still highly debated and defended as a possible principle driving force of plate tectonics. The other forces are only used in global geodynamic models not using plate tectonics concepts (therefore beyond the discussions treated in this section) or proposed as minor modulations within the overall plate tectonics model.

In 1973, George W. Moore of the USGS and R. C. Bostrom presented evidence for a general westward drift of the Earth's lithosphere with respect to the mantle. He concluded that tidal forces (the tidal lag or "friction") caused by the Earth's rotation and the forces acting upon it by the Moon are a driving force for plate tectonics. As the Earth spins eastward beneath the moon, the moon's gravity ever so slightly pulls the Earth's surface layer back westward, just as proposed by Alfred Wegener. In a more recent 2006 study, scientists reviewed and advocated these earlier proposed ideas. It has also been suggested recently in Lovett (2006) that this observation may also explain why Venus and Mars have no plate tectonics, as Venus has no moon and Mars' moons are too small to have significant tidal effects on the planet. In a recent paper, it was suggested that, on the other hand, it can easily be observed that many plates are moving north and eastward, and that the dominantly westward motion of the Pacific ocean basins derives simply from the eastward bias of the Pacific spreading center (which is not a predicted manifestation of such lunar forces). In the same paper the authors admit, however, that relative to the lower mantle, there is a slight westward component in the motions of all the plates. They demonstrated though that the westward drift, seen only for the past 30 Ma, is attributed to the increased dominance of the steadily growing and accelerating Pacific plate. The debate is still open.

Relative Significance of Each Driving Force Mechanism

The vector of a plate's motion is a function of all the forces acting on the plate; however, therein lies the problem regarding the degree to which each process contributes to the overall motion of each tectonic plate.

The diversity of geodynamic settings and the properties of each plate result from the impact of the various processes actively driving each individual plate. One method of dealing with this problem is to consider the relative rate at which each plate is moving as well as the evidence related to the significance of each process to the overall driving force on the plate.

One of the most significant correlations discovered to date is that lithospheric plates attached to downgoing (subducting) plates move much faster than plates not attached to subducting plates. The Pacific plate, for instance, is essentially surrounded by zones of subduction (the so-called Ring of Fire) and moves much faster than the plates of the Atlantic basin, which are attached (perhaps one could say 'welded') to adjacent continents instead of subducting plates. It is thus thought that forces associated with the downgoing plate (slab pull and slab suction) are the driving forces which determine the motion of plates, except for those plates which are not being subducted. The driving forces of plate motion continue to be active subjects of on-going research within geophysics and tectonophysics.

Development of the Theory

Summary

Detailed map showing the tectonic plates with their movement vectors.

In line with other previous and contemporaneous proposals, in 1912 the meteorologist Alfred Wegener amply described what he called continental drift, expanded in his 1915 book *The Origin of Continents and Oceans* and the scientific debate started that would end up fifty years later in the theory of plate tectonics. Starting from the idea (also expressed by his forerunners) that the present continents once formed a single land mass (which was called Pangea later on) that drifted apart, thus releasing the continents from the Earth's mantle and likening them to "icebergs" of low density granite floating on a sea of denser basalt. Supporting evidence for the idea came from the dove-tailing outlines of South America's east coast and Africa's west coast, and from the matching of the rock formations along these edges. Confirmation of their previous contiguous nature also came from the fossil plants *Glossopteris* and *Gangamopteris*, and the therapsid or mammal-like reptile *Lystrosaurus*, all widely distributed over South America, Africa, Antarctica, India and Australia. The evidence for such an erstwhile joining of these continents was patent to field geologists working in the southern hemisphere. The South African Alex du Toit put together a mass of such information in his 1937 publication *Our Wan-*

dering Continents, and went further than Wegener in recognising the strong links between the Gondwana fragments.

But without detailed evidence and a force sufficient to drive the movement, the theory was not generally accepted: the Earth might have a solid crust and mantle and a liquid core, but there seemed to be no way that portions of the crust could move around. Distinguished scientists, such as Harold Jeffreys and Charles Schuchert, were outspoken critics of continental drift.

Despite much opposition, the view of continental drift gained support and a lively debate started between "drifters" or "mobilists" (proponents of the theory) and "fixists" (opponents). During the 1920s, 1930s and 1940s, the former reached important milestones proposing that convection currents might have driven the plate movements, and that spreading may have occurred below the sea within the oceanic crust. Concepts close to the elements now incorporated in plate tectonics were proposed by geophysicists and geologists (both fixists and mobilists) like Vening-Meinesz, Holmes, and Umbgrove.

One of the first pieces of geophysical evidence that was used to support the movement of lithospheric plates came from paleomagnetism. This is based on the fact that rocks of different ages show a variable magnetic field direction, evidenced by studies since the mid–nineteenth century. The magnetic north and south poles reverse through time, and, especially important in paleotectonic studies, the relative position of the magnetic north pole varies through time. Initially, during the first half of the twentieth century, the latter phenomenon was explained by introducing what was called "polar wander", i.e., it was assumed that the north pole location had been shifting through time. An alternative explanation, though, was that the continents had moved (shifted and rotated) relative to the north pole, and each continent, in fact, shows its own "polar wander path". During the late 1950s it was successfully shown on two occasions that these data could show the validity of continental drift: by Keith Runcorn in a paper in 1956, and by Warren Carey in a symposium held in March 1956.

The second piece of evidence in support of continental drift came during the late 1950s and early 60s from data on the bathymetry of the deep ocean floors and the nature of the oceanic crust such as magnetic properties and, more generally, with the development of marine geology which gave evidence for the association of seafloor spreading along the mid-oceanic ridges and magnetic field reversals, published between 1959 and 1963 by Heezen, Dietz, Hess, Mason, Vine & Matthews, and Morley.

Simultaneous advances in early seismic imaging techniques in and around Wadati-Benioff zones along the trenches bounding many continental margins, together with many other geophysical (e.g. gravimetric) and geological observations, showed how the oceanic crust could disappear into the mantle, providing the mechanism to balance the extension of the ocean basins with shortening along its margins.

All this evidence, both from the ocean floor and from the continental margins, made it clear around 1965 that continental drift was feasible and the theory of plate tectonics, which was defined in a series of papers between 1965 and 1967, was born, with all its extraordinary explanatory and predictive power. The theory revolutionized the Earth sciences, explaining a diverse range of geological phenomena and their implications in other studies such as paleogeography and paleobiology.

Continental Drift

In the late 19th and early 20th centuries, geologists assumed that the Earth's major features were fixed, and that most geologic features such as basin development and mountain ranges could be explained by vertical crustal movement, described in what is called the geosynclinal theory. Generally, this was placed in the context of a contracting planet Earth due to heat loss in the course of a relatively short geological time.

Alfred Wegener in Greenland in the winter of 1912-13.

It was observed as early as 1596 that the opposite coasts of the Atlantic Ocean—or, more precisely, the edges of the continental shelves—have similar shapes and seem to have once fitted together.

Since that time many theories were proposed to explain this apparent complementarity, but the assumption of a solid Earth made these various proposals difficult to accept.

The discovery of radioactivity and its associated heating properties in 1895 prompted a re-examination of the apparent age of the Earth. This had previously been estimated by its cooling rate and assumption the Earth's surface radiated like a black body. Those calculations had implied that, even if it started at red heat, the Earth would have dropped to its present temperature in a few tens of millions of years. Armed with the knowledge of a new heat source, scientists realized that the Earth would be much older, and that its core was still sufficiently hot to be liquid.

By 1915, after having published a first article in 1912, Alfred Wegener was making serious arguments for the idea of continental drift in the first edition of *The Origin of Continents and Oceans*. In that book (re-issued in four successive editions up to the final one in 1936), he noted how the east coast of South America and the west coast of Africa looked as if they were once attached. Wegener was not the first to note this (Abraham Ortelius, Antonio Snider-Pellegrini, Eduard Suess, Roberto Mantovani and Frank Bursley Taylor preceded him just to mention a few), but he was the first to marshal significant fossil and paleo-topographical and climatological evidence to support this simple observation (and was supported in this by researchers such as Alex du Toit). Furthermore, when the rock strata of the margins of separate continents are very similar it suggests that these rocks were formed in the same way, implying that they were joined initially. For instance, parts of Scotland and Ireland contain rocks very similar to those found in Newfoundland and New Brunswick. Furthermore, the Caledonian Mountains of Europe and parts of the Appalachian Mountains of North America are very similar in structure and lithology.

However, his ideas were not taken seriously by many geologists, who pointed out that there was no apparent mechanism for continental drift. Specifically, they did not see how continental rock could plow through the much denser rock that makes up oceanic crust. Wegener could not explain the force that drove continental drift, and his vindication did not come until after his death in 1930.

Floating Continents, Paleomagnetism, and Seismicity Zones

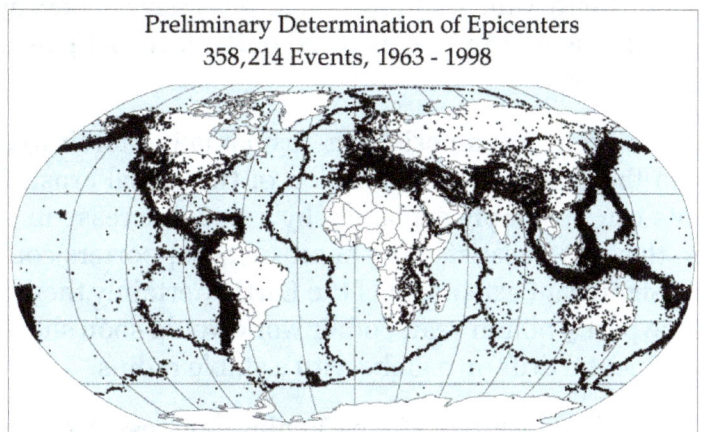

Global earthquake epicenters, 1963–1998

As it was observed early that although granite existed on continents, seafloor seemed to be composed of denser basalt, the prevailing concept during the first half of the twentieth century was that there were two types of crust, named "sial" (continental type crust) and "sima" (oceanic type crust). Furthermore, it was supposed that a static shell of strata was present under the continents. It therefore looked apparent that a layer of basalt (sial) underlies the continental rocks.

However, based on abnormalities in plumb line deflection by the Andes in Peru, Pierre Bouguer had deduced that less-dense mountains must have a downward projection into the denser layer underneath. The concept that mountains had "roots" was confirmed by George B. Airy a hundred years later, during study of Himalayan gravitation, and seismic studies detected corresponding density variations. Therefore, by the mid-1950s, the question remained unresolved as to whether mountain roots were clenched in surrounding basalt or were floating on it like an iceberg.

During the 20th century, improvements in and greater use of seismic instruments such as seismographs enabled scientists to learn that earthquakes tend to be concentrated in specific areas, most notably along the oceanic trenches and spreading ridges. By the late 1920s, seismologists were beginning to identify several prominent earthquake zones parallel to the trenches that typically were inclined 40–60° from the horizontal and extended several hundred kilometers into the Earth. These zones later became known as Wadati-Benioff zones, or simply Benioff zones, in honor of the seismologists who first recognized them, Kiyoo Wadati of Japan and Hugo Benioff of the United States. The study of global seismicity greatly advanced in the 1960s with the establishment of the Worldwide Standardized Seismograph Network (WWSSN) to monitor the compliance of the 1963 treaty banning above-ground testing of nuclear weapons. The much improved data from the

WWSSN instruments allowed seismologists to map precisely the zones of earthquake concentration worldwide.

Meanwhile, debates developed around the phenomena of polar wander. Since the early debates of continental drift, scientists had discussed and used evidence that polar drift had occurred because continents seemed to have moved through different climatic zones during the past. Furthermore, paleomagnetic data had shown that the magnetic pole had also shifted during time. Reasoning in an opposite way, the continents might have shifted and rotated, while the pole remained relatively fixed. The first time the evidence of magnetic polar wander was used to support the movements of continents was in a paper by Keith Runcorn in 1956, and successive papers by him and his students Ted Irving (who was actually the first to be convinced of the fact that paleomagnetism supported continental drift) and Ken Creer.

This was immediately followed by a symposium in Tasmania in March 1956. In this symposium, the evidence was used in the theory of an expansion of the global crust. In this hypothesis the shifting of the continents can be simply explained by a large increase in size of the Earth since its formation. However, this was unsatisfactory because its supporters could offer no convincing mechanism to produce a significant expansion of the Earth. Certainly there is no evidence that the moon has expanded in the past 3 billion years; other work would soon show that the evidence was equally in support of continental drift on a globe with a stable radius.

During the thirties up to the late fifties, works by Vening-Meinesz, Holmes, Umbgrove, and numerous others outlined concepts that were close or nearly identical to modern plate tectonics theory. In particular, the English geologist Arthur Holmes proposed in 1920 that plate junctions might lie beneath the sea, and in 1928 that convection currents within the mantle might be the driving force. Often, these contributions are forgotten because:

- At the time, continental drift was not accepted.

- Some of these ideas were discussed in the context of abandoned fixistic ideas of a deforming globe without continental drift or an expanding Earth.

- They were published during an episode of extreme political and economic instability that hampered scientific communication.

- Many were published by European scientists and at first not mentioned or given little credit in the papers on sea floor spreading published by the American researchers in the 1960s.

Mid-oceanic Ridge Spreading and Convection

In 1947, a team of scientists led by Maurice Ewing utilizing the Woods Hole Oceanographic Institution's research vessel *Atlantis* and an array of instruments, confirmed the existence of a rise in the central Atlantic Ocean, and found that the floor of the seabed beneath the layer of sediments consisted of basalt, not the granite which is the main constituent of continents. They also found that the oceanic crust was much thinner than continental crust. All these new findings raised important and intriguing questions.

The new data that had been collected on the ocean basins also showed particular characteristics regarding the bathymetry. One of the major outcomes of these datasets was that all along the globe, a system of mid-oceanic ridges was detected. An important conclusion was that along this system, new ocean floor was being created, which led to the concept of the "Great Global Rift". This was described in the crucial paper of Bruce Heezen (1960), which would trigger a real revolution in thinking. A profound consequence of seafloor spreading is that new crust was, and still is, being continually created along the oceanic ridges. Therefore, Heezen advocated the so-called "expanding Earth" hypothesis of S. Warren Carey. So, still the question remained: how can new crust be continuously added along the oceanic ridges without increasing the size of the Earth? In reality, this question had been solved already by numerous scientists during the forties and the fifties, like Arthur Holmes, Vening-Meinesz, Coates and many others: The crust in excess disappeared along what were called the oceanic trenches, where so-called "subduction" occurred. Therefore, when various scientists during the early sixties started to reason on the data at their disposal regarding the ocean floor, the pieces of the theory quickly fell into place.

The question particularly intrigued Harry Hammond Hess, a Princeton University geologist and a Naval Reserve Rear Admiral, and Robert S. Dietz, a scientist with the U.S. Coast and Geodetic Survey who first coined the term *seafloor spreading*. Dietz and Hess (the former published the same idea one year earlier in *Nature*, but priority belongs to Hess who had already distributed an unpublished manuscript of his 1962 article by 1960) were among the small handful who really understood the broad implications of sea floor spreading and how it would eventually agree with the, at that time, unconventional and unaccepted ideas of continental drift and the elegant and mobilistic models proposed by previous workers like Holmes.

In the same year, Robert R. Coats of the U.S. Geological Survey described the main features of island arc subduction in the Aleutian Islands. His paper, though little noted (and even ridiculed) at the time, has since been called "seminal" and "prescient". In reality, it actually shows that the work by the European scientists on island arcs and mountain belts performed and published during the 1930s up until the 1950s was applied and appreciated also in the United States.

If the Earth's crust was expanding along the oceanic ridges, Hess and Dietz reasoned like Holmes and others before them, it must be shrinking elsewhere. Hess followed Heezen, suggesting that new oceanic crust continuously spreads away from the ridges in a conveyor belt–like motion. And, using the mobilistic concepts developed before, he correctly concluded that many millions of years later, the oceanic crust eventually descends along the continental margins where oceanic trenches – very deep, narrow canyons – are formed, e.g. along the rim of the Pacific Ocean basin. The important step Hess made was that convection currents would be the driving force in this process, arriving at the same conclusions as Holmes had decades before with the only difference that the thinning of the ocean crust was performed using Heezen's mechanism of spreading along the ridges. Hess therefore concluded that the Atlantic Ocean was expanding while the Pacific Ocean was shrinking. As old oceanic crust is "consumed" in the trenches (like Holmes and others, he thought this was done by thickening of the continental lithosphere, not, as now understood, by underthrusting at a larger scale of the oceanic crust itself into the mantle), new magma rises and erupts along the spreading ridges to form new crust. In effect, the ocean basins are perpetually being "recycled," with the creation of new crust and the destruction of old oceanic lithosphere occurring simultaneously. Thus, the new mobilistic concepts neatly explained why the Earth does not get bigger with sea floor spreading, why there is so little sediment accumulation on the ocean floor, and why oceanic rocks are much younger than continental rocks.

Magnetic Striping

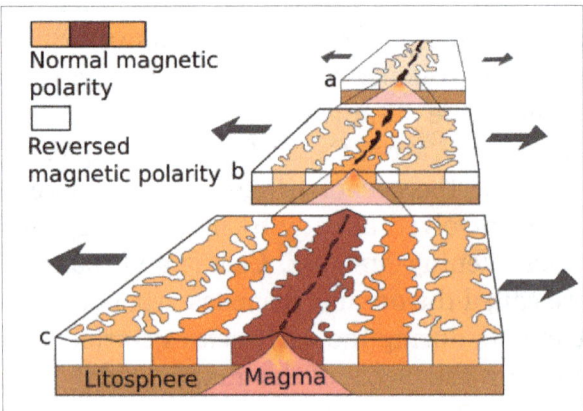

Seafloor magnetic striping.

Beginning in the 1950s, scientists like Victor Vacquier, using magnetic instruments (magnetometers) adapted from airborne devices developed during World War II to detect submarines, began recognizing odd magnetic variations across the ocean floor. This finding, though unexpected, was not entirely surprising because it was known that basalt—the iron-rich, volcanic rock making up the ocean floor—contains a strongly magnetic mineral (magnetite) and can locally distort compass readings. This distortion was recognized by Icelandic mariners as early as the late 18th century. More important, because the presence of magnetite gives the basalt measurable magnetic properties, these newly discovered magnetic variations provided another means to study the deep ocean floor. When newly formed rock cools, such magnetic materials recorded the Earth's magnetic field at the time.

A demonstration of magnetic striping. (The darker the color is, the closer it is to normal polarity).

As more and more of the seafloor was mapped during the 1950s, the magnetic variations turned out not to be random or isolated occurrences, but instead revealed recognizable patterns. When these magnetic patterns were mapped over a wide region, the ocean floor showed a zebra-like pattern: one stripe with normal polarity and the adjoining stripe with reversed polarity. The overall pattern, defined by these alternating bands of normally and reversely polarized rock, became known as magnetic striping, and was published by Ron G. Mason and co-workers in 1961, who did not find, though, an explanation for these data in terms of sea floor spreading, like Vine, Matthews and Morley a few years later.

The discovery of magnetic striping called for an explanation. In the early 1960s scientists such as Heezen, Hess and Dietz had begun to theorise that mid-ocean ridges mark structurally weak zones where the ocean floor was being ripped in two lengthwise along the ridge crest. New magma from deep within the Earth rises easily through these weak zones and eventually erupts along the crest of the ridges to create new oceanic crust. This process, at first denominated the "conveyer belt hypothesis" and later called seafloor spreading, operating over many millions of years continues to form new ocean floor all across the 50,000 km-long system of mid-ocean ridges.

Only four years after the maps with the "zebra pattern" of magnetic stripes were published, the link between sea floor spreading and these patterns was correctly placed, independently by Lawrence Morley, and by Fred Vine and Drummond Matthews, in 1963, now called the Vine-Matthews-Morley hypothesis. This hypothesis linked these patterns to geomagnetic reversals and was supported by several lines of evidence:

- The stripes are symmetrical around the crests of the mid-ocean ridges; at or near the crest of the ridge, the rocks are very young, and they become progressively older away from the ridge crest;

- The youngest rocks at the ridge crest always have present-day (normal) polarity;

- Stripes of rock parallel to the ridge crest alternate in magnetic polarity (normal-reversed-normal, etc.), suggesting that they were formed during different epochs documenting the (already known from independent studies) normal and reversal episodes of the Earth's magnetic field.

By explaining both the zebra-like magnetic striping and the construction of the mid-ocean ridge system, the seafloor spreading hypothesis (SFS) quickly gained converts and represented another major advance in the development of the plate-tectonics theory. Furthermore, the oceanic crust now came to be appreciated as a natural "tape recording" of the history of the geomagnetic field reversals (GMFR) of the Earth's magnetic field. Today, extensive studies are dedicated to the calibration of the normal-reversal patterns in the oceanic crust on one hand and known timescales derived from the dating of basalt layers in sedimentary sequences (magnetostratigraphy) on the other, to arrive at estimates of past spreading rates and plate reconstructions.

Definition and Refining of the Theory

After all these considerations, Plate Tectonics (or, as it was initially called "New Global Tectonics") became quickly accepted in the scientific world, and numerous papers followed that defined the concepts:

- In 1965, Tuzo Wilson who had been a promotor of the sea floor spreading hypothesis and continental drift from the very beginning added the concept of transform faults to the model, completing the classes of fault types necessary to make the mobility of the plates on the globe work out.

- A symposium on continental drift was held at the Royal Society of London in 1965 which must be regarded as the official start of the acceptance of plate tectonics by the scientific community, and which abstracts are issued as Blacket, Bullard & Runcorn (1965). In this symposium, Edward Bullard and co-workers showed with a computer calculation how the continents along both sides of the Atlantic would best fit to close the ocean, which became known as the famous "Bullard's Fit".

- In 1966 Wilson published the paper that referred to previous plate tectonic reconstructions, introducing the concept of what is now known as the "Wilson Cycle".

- In 1967, at the American Geophysical Union's meeting, W. Jason Morgan proposed that the Earth's surface consists of 12 rigid plates that move relative to each other.

- Two months later, Xavier Le Pichon published a complete model based on 6 major plates with their relative motions, which marked the final acceptance by the scientific community of plate tectonics.

- In the same year, McKenzie and Parker independently presented a model similar to Morgan's using translations and rotations on a sphere to define the plate motions.

Implications for Biogeography

Continental drift theory helps biogeographers to explain the disjunct biogeographic distribution of present-day life found on different continents but having similar ancestors. In particular, it explains the Gondwanan distribution of ratites and the Antarctic flora.

Plate Reconstruction

Reconstruction is used to establish past (and future) plate configurations, helping determine the shape and make-up of ancient supercontinents and providing a basis for paleogeography.

Defining Plate Boundaries

Current plate boundaries are defined by their seismicity. Past plate boundaries within existing plates are identified from a variety of evidence, such as the presence of ophiolites that are indicative of vanished oceans.

Past Plate Motions

Tectonic Motion First Began Around Three Billion Years Ago.

Various types of quantitative and semi-quantitative information are available to constrain past plate motions. The geometric fit between continents, such as between west Africa and South America is still an important part of plate reconstruction. Magnetic stripe patterns provide a reliable guide to relative plate motions going back into the Jurassic period. The tracks of hotspots give absolute reconstructions, but these are only available back to the Cretaceous. Older reconstructions rely mainly on paleomagnetic pole data, although these only constrain the latitude and rotation, but not the longitude. Combining poles of different ages in a particular plate to produce apparent polar wander paths provides a method for comparing the motions of different plates through time. Additional evidence comes from the distribution of certain sedimentary rock types, faunal provinces shown by particular fossil groups, and the position of orogenic belts.

Formation and Break-up of Continents

The movement of plates has caused the formation and break-up of continents over time, including occasional formation of a supercontinent that contains most or all of the continents. The supercontinent Columbia or Nuna formed during a period of 2,000 to 1,800 million years ago and broke up about 1,500 to 1,300 million years ago. The supercontinent Rodinia is thought to have formed about 1 billion years ago and to have embodied most or all of Earth's continents, and broken up into eight continents around 600 million years ago. The eight continents later re-assembled into another supercontinent called Pangaea; Pangaea broke up into Laurasia (which became North America and Eurasia) and Gondwana (which became the remaining continents).

The Himalayas, the world's tallest mountain range, are assumed to have been formed by the collision of two major plates. Before uplift, they were covered by the Tethys Ocean.

Current Plates

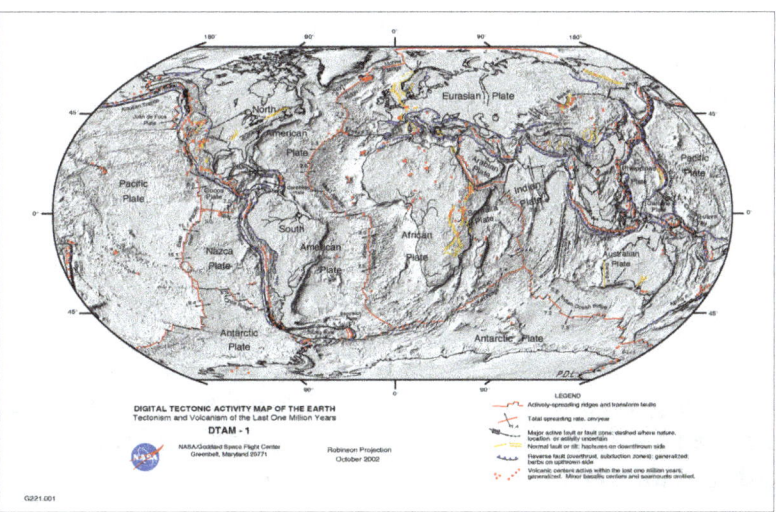

Depending on how they are defined, there are usually seven or eight "major" plates: African, Antarctic, Eurasian, North American, South American, Pacific, and Indo-Australian. The latter is sometimes subdivided into the Indian and Australian plates.

There are dozens of smaller plates, the seven largest of which are the Arabian, Caribbean, Juan de Fuca, Cocos, Nazca, Philippine Sea and Scotia.

The current motion of the tectonic plates is today determined by remote sensing satellite data sets, calibrated with ground station measurements.

Other Celestial Bodies (Planets, Moons)

The appearance of plate tectonics on terrestrial planets is related to planetary mass, with more massive planets than Earth expected to exhibit plate tectonics. Earth may be a borderline case, owing its tectonic activity to abundant water (silica and water form a deep eutectic.)

Venus

Venus shows no evidence of active plate tectonics. There is debatable evidence of active tectonics in the planet's distant past; however, events taking place since then (such as the plausible and generally accepted hypothesis that the Venusian lithosphere has thickened greatly over the course of several hundred million years) has made constraining the course of its geologic record difficult. However, the numerous well-preserved impact craters have been utilized as a dating method to approximately date the Venusian surface (since there are thus far no known samples of Venusian rock to be dated by more reliable methods). Dates derived are dominantly in the range 500 to 750 million years ago, although ages of up to 1,200 million years ago have been calculated. This research has led to the fairly well accepted hypothesis that Venus has undergone an essentially complete volcanic resurfacing at least once in its distant past, with the last event taking place approximately within the range of estimated surface ages. While the mechanism of such an impressive thermal event remains a debated issue in Venusian geosciences, some scientists are advocates of processes involving plate motion to some extent.

One explanation for Venus's lack of plate tectonics is that on Venus temperatures are too high for significant water to be present. The Earth's crust is soaked with water, and water plays an important role in the development of shear zones. Plate tectonics requires weak surfaces in the crust along which crustal slices can move, and it may well be that such weakening never took place on Venus because of the absence of water. However, some researchers[who?] remain convinced that plate tectonics is or was once active on this planet.

Mars

Mars is considerably smaller than Earth and Venus, and there is evidence for ice on its surface and in its crust.

In the 1990s, it was proposed that Martian Crustal Dichotomy was created by plate tectonic processes. Scientists today disagree, and think that it was created either by upwelling within the Martian mantle that thickened the crust of the Southern Highlands and formed Tharsis or by a giant impact that excavated the Northern Lowlands.

Valles Marineris may be a tectonic boundary.

Observations made of the magnetic field of Mars by the *Mars Global Surveyor* spacecraft in 1999 showed patterns of magnetic striping discovered on this planet. Some scientists interpreted these as requiring plate tectonic processes, such as seafloor spreading. However, their data fail a "magnetic reversal test", which is used to see if they were formed by flipping polarities of a global magnetic field.

Icy Satellites

Some of the satellites of Jupiter have features that may be related to plate-tectonic style deformation, although the materials and specific mechanisms may be different from plate-tectonic activity on Earth. On 8 September 2014, NASA reported finding evidence of plate tectonics on Europa, a satellite of Jupiter—the first sign of such geological activity on another world other than Earth.

Titan, the largest moon of Saturn, was reported to show tectonic activity in images taken by the *Huygens* probe, which landed on Titan on January 14, 2005.

Exoplanets

On Earth-sized planets, plate tectonics is more likely if there are oceans of water; however, in 2007, two independent teams of researchers came to opposing conclusions about the likelihood of plate tectonics on larger super-earths with one team saying that plate tectonics would be episodic or stagnant and the other team saying that plate tectonics is very likely on super-earths even if the planet is dry.

Ocean Turbidity

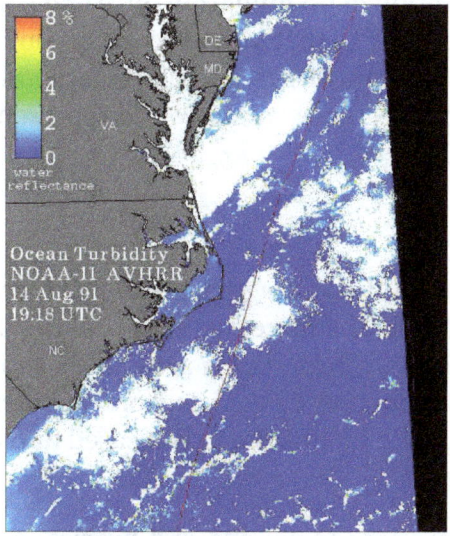

Visualisation of the Ocean Turbidity of the ocean just before Hurricane Bob (August 14, 1991).

Ocean turbidity is a measure of the amount of cloudiness or haziness in sea water caused by individual particles that are too small to be seen without magnification. Highly turbid ocean waters are those with a large number of scattering particulates in them. In both highly absorbing and highly scattering waters, visibility into the water is reduced. The highly scattering (turbid) water still reflects a lot of light while the highly absorbing water, such as a blackwater river or lake, is very dark. The scattering particles that cause the water to be turbid can be composed of many things, including sediments and phytoplankton.

Measurement

There are a number of ways to measure ocean turbidity, including autonomous remote vehicles, shipcasts and satellites.

From a satellite, a proxy measurement of the water turbidity can be made by examining the amount of reflectance in the visible region of the electromagnetic spectrum. For the Advanced Very High Resolution Radiometer (AVHRR), the logical choice is band 1, covering wavelengths 580 to 680 nanometres, the orange and red. In order to make derived products that are comparable over time and space, an atmospheric correction is required. To do this, the effects of Rayleigh scattering are calculated based on the satellite viewing angle and the solar zenith angle and then subtracted from the band 1 radiance. For an aerosol correction, band 2 in the near infrared is used. It is first corrected for Rayleigh scattering and then subtracted from the Rayleigh corrected band 1. The Rayleigh corrected band 2 is assumed to be aerosol radiance because no return signal from water in the near infrared is expected since water is highly absorbing at those wavelengths. Because bands 1 and 2 are relatively close on the electromagnetic spectrum, we can reasonably assume their aerosol radiances are the same.

In these images the turbidity is quantified as the percent reflected light emerging from the water column in a range of 0 to 8 percent. The reflectance percentage can be correlated to attenuation, Secchi disk depth or total suspended solids although the exact relationship will vary regionally and depends on the optical properties of the water. For example, in Florida Bay, 10% reflectance corresponds to a sediment concentration of 30 milligram/litre and a Secchi depth of 0.5 metre. These relationships are approximately linear so that 5% reflectance would correspond to a sediment concentration of approximately 15 milligram/litre and a Secchi depth of 1 metre. In the Mississippi River plume regions these same reflectance values would represent sediment concentrations that are about ten times or more higher.

Hurricanes

As one would expect, the majority of these images reveal large increases in turbidity in the regions where a hurricane has made landfall. The increases are primarily due to sediments that have been resuspended from the shallow bottom regions. In areas near shore some of the signal may also be due to sediments eroded from beaches as well as from sediment laden river plumes. In some cases a post-hurricane phytoplankton bloom due to increased nutrient availability may perhaps be detectable.

The examination of the turbidity after the passing of a hurricane can have potentially many uses for coastal resource management including:

- Identifying regional "hot spots" where the erosion could be expected to be most severe.

- Estimating the total sediment concentration that has been mobilized by the hurricane.

- Determining the spatial extent of the sediment mobilization.

- Identifying the extent and contribution of river plumes.

- Assessing and predicting potential ecosystem impacts.

With regard to these uses, determining the regions of high turbidity will allow managers to best decide on response strategies as well as help ensure that post-hurricane resources are most effectively utilized.

Interpreting Images

Only a small fraction of the light incident on the ocean will be reflected and received by the satellite. The probability for a photon to reflect and exit the ocean decreases exponentially with length of its path through the water because the ocean is an absorbing medium. The more ocean a photon must travel through, the greater its chances of being absorbed by something. After absorption, it will eventually become part of the ocean's heat reservoir. The absorption and scattering characteristics of a water body determine the rate of vertical light attenuation and set a limit to the depths contributing to a satellite signal. A reasonable rule of thumb is that 90 percent of the signal coming from the water that is seen by the satellite is from the first attenuation length. How deep this is depends on the absorption and scattering properties of both the water itself and other constituents in the water. For wavelengths in the near infrared and longer, the penetration depth varies from a metre to a few micrometres. For band 1, the penetration depth will usually be between 1 and 10 metres. If the water has a large turbidity spike below 10 metres, the spike is unlikely to be seen by a satellite.

For very shallow clear water there is a good chance the bottom may be seen. For example, in the Bahamas, the water is quite clear and only a few metres deep, resulting in an apparent high turbidity because the bottom reflects a lot of the band 1 light. For areas with consistently high turbidity signals, particularly areas with relatively clear water, part of the signal may be due to bottom reflection. Normally this will not be a problem with a post-hurricane turbidity image since the storm easily resuspends enough sediment such that bottom reflection is negligible.

Clouds are also problematic for the interpretation of satellite derived turbidity. Cloud removal algorithms perform a satisfactory job for pixels that are fully cloudy. Partially cloudy pixels are much harder to identify and typically result in false high turbidity estimates. High turbidity values near clouds are suspect.

References

- Richard Stenger (September 19, 2000). "Flotilla of sensors to monitor world's oceans". CNN. Archived from the original on 6 November 2007. Retrieved 19, January 2020

- "Argo Floats : How do we measure the ocean?" (Youtube video). Integrated Marine Observing Strategy. 10 March 2014. Retrieved 01, March 2020

- Davidson, Helen (30 January 2014). "Scientists to launch bio robots in Indian Ocean to study its 'interior biology'". The Guardian. Retrieved 19, February 2020

- Kenneth Chang (June 24, 1999). "An Ear to Ocean Temperature". ABC News. Archived from the original on 2003-10-06. Retrieved 01, April 2020

- Frankel, A. S.; C. W. Clark (2002). "ATOC and other factors affecting distribution and abundance of humpback whales (Megaptera novaeangliae) off the north shore of Kauai". 18. Marine Mammal Science. pp. 664–662. Retrieved 29, August 2020

- National Research Council (2000). Marine mammals and low-frequency sound: Progress since 1994. Washington, D.C.: National Academy Press

- Frankel, A. S.; C. W. Clark (2000). "Behavioral responses of humpback whales (Megaptera novaeangliae) to full-scale ATOC signals". 108. Journal of the Acoustical Society of America. pp. 1–8. Retrieved 17, April 2020

- Potter, J. R. (1994). "ATOC: Sound Policy or Enviro-Vandalism? Aspects of a Modern Media-Fueled Policy Issue". 3. The Journal of Environment & Development. pp. 47–62. doi:10.1177/107049659400300205. Retrieved 28, February 2020

- Bows, Kevin; Bows, Alice (2011). "Beyond 'dangerous' climate change: emission scenarios for a new world" (PDF). Philosophical Transactions of the Royal Society A. 369 (1934): 20–44. Bibcode:2011RSPTA.369...20A. doi:10.1098/rsta.2010.0290. Retrieved 14, August 2020

- Bombosch, A. (2014). "Predictive habitat modelling of humpback (Megaptera novaeangliae) and Antarctic minke (Balaenoptera bonaerensis) whales in the Southern Ocean as a planning tool for seismic surveys".

Deep-Sea Research Part I: Oceanographic Research Papers. Deep-Sea Research Part I. 91: 101–114. Bibcode:2014DSRI...91..101B. doi:10.1016/j.dsr.2014.05.017. Retrieved 24, April 2020

• Garrison, Tom (2009). Oceanography: An Invitation to Marine Science (7th ed.). Cengage Learning. p. 582. ISBN 9780495391937

• National Research Council (2003). Ocean Noise and Marine Mammals. National Academies Press. ISBN 978-0-309-08536-6. Retrieved 01, June 2020

• Kump, Lee R.; Kasting, James F.; Crane, Robert G. (2003). The Earth System (2nd ed.). Upper Saddle River: Prentice Hall. pp. 162–164. ISBN 0-613-91814-2

• Mobley, J. R. (2005). "Assessing responses of humpback whales to North Pacific Acoustic Laboratory (NPAL) transmissions: Results of 2001-2003 aerial surveys north of Kauai". 117. Journal of the Acoustical Society of America. pp. 1666–1673. Retrieved 19, March 2020

Chapter 6

Ocean Observations and its Model

Ocean observations involve monitoring various aspects of the ocean such as air temperature, precipitation, surface wind stress, water vapor, sea surface temperature, sea surface salinity, backscatter, turbidity, etc. This chapter closely examines ocean observations as well as the ocean reanalysis method to provide an extensive understanding of the subject.

Ocean Observations

The following are considered essential ocean climate variables by the Ocean Observations Panel for Climate (OOPC) that are currently feasible with current observational systems.

Ocean Climate Variables

Atmosphere Surface

- Air Temperature
- Precipitation (meteorology)
- evaporation
- Air Pressure, sea level pressure (SLP)
- Surface radiative fluxes
- Surface thermodynamic fluxes
- Wind speed and direction
- Surface wind stress
- Water vapor

Ocean Surface

- Sea surface temperature (SST)
- Sea surface salinity (SSS)
- Sea level
- Sea state
- Sea ice

- Ocean current

- Ocean color (for biological activity)

- Carbon dioxide partial pressure (pCO2)

Ocean Subsurface

- Backscatter

- Carbon Dioxide

- Chlorophyll

- Conductivity

- Density

- Iron

- Irradiance

- Nutrients

  o Nitrate

- Methane

- Ocean current

  o Single Point

  o Water Column

- Ocean tracers

- Oxygen

- Phytoplankton

- Salinity

- Sigma-T

- Sound Velocity

- Temperature

- Turbidity

Ocean Observation Sources

Satellite

There is a composite network of satellites that generate observations. These include:

Type	Variables observed	Responsible organizations
Infrared (IR)	SST, sea ice	CEOS, IGOS, CGMS
AMSR-class microwave	SST, wind speed, sea ice	CEOS, IGOS, CGMS
Surface vector wind (two wide-swath scatterometers desired)	surface vector wind, sea ice	CEOS, IGOS, CGMS
Ocean color	chlorophyll concentration (biomass of phytoplankton)	IOCCG
high-precision altimetry	sea-level anomaly from steady state	CEOS, IGOS, CGMS
low-precision altimetry	sea level	CEOS, IGOS, CGMS
Synthetic aperture radar	sea ice, sea state	CEOS, IGOS, CGMS

In Situ

There is a composite network of in situ observations. These include:

Type	Variables observed	Responsible organizations
Global surface drifting buoy array with 5 degree resolution (1250 total)	SST, SLP, Current (based on position change)	JCOMM Data Buoy Cooperation Panel (DBCP)
Global tropical moored buoy network (about 120 moorings)	typically SST and surface vector wind, but can also include SLP, current, air-sea flux variables	JCOMM DBCP Tropical Moored Buoy Implementation Panel (TIP)
Volunteer Observing Ship (VOS) fleet	all feasible surface ECVs	JCOMM Ship Observations Team (SOT)
VOSClim	all feasible surface ECVs plus extensive ship metadata	JCOMM Ship Observations Team (SOT)
Global referencing mooring network (29 moorings)	all feasible surface ECVs	OceanSITES
GLOSS core sea-level network, plus regional/national networks	sea level	JCOMM GLOSS
Carbon VOS	pCO2, SST, SSS	IOCCP
Sea ice buoys	sea ice	JCOMM DBCP IABP and IPAB

Subsurface

There is a composite network of subsurface observations. These include:

Type	Variables observed	Responsible organizations
Repeat XBT (Expendable bathythermograph) line network (41 lines)	Temperature	JCOMM Ship Observations Team (SOT)

Global tropical moored buoy network (~120 moorings)	Temperature, salinity, current, other feasible autonomously observable ECVs	JCOMM DBCP Tropical Moored Buoy Implementation Panel (TIP)
Reference mooring network (29 moorings)	all autonomously observable ECVs	OceanSITES
Sustained and repeated ship-based hydrography network	All feasible ECVs, including those that depend on obtaining water samples	IOCCP, CLIVAR, other national efforts
Argo (oceanography) network	temperature, salinity, current	Argo
Critical current and transport monitoring	temperature, heat, freshwater, carbon transports, mass	CLIVAR, IOCCP, OceanSITES
Regional and global synthesis programmes	inferred currents, transports gridded fields of all ECVs	GODAE, CLIVAR, other national efforts
Cabled ocean observatories	audio, backscatter, chlorophyll, CO_2, conductivity, currents, density, Eh, gravity, iron, irradiance, methane, nitrate, oxygen, pressure, salinity, seismic, sigma-T, sound velocity, temperature, turbidity, video	Ocean Networks Canada, Monterey Accelerated Research System, Ocean Observatories Initiative, ALOHA, ESONET (European Seas Observatory NETwork), Dense Oceanfloor Network System for Earthquakes and Tsunamis (DONET), Fixed-Point Open Ocean Observatories (FixO3).

Accuracy of Measurements

The quality of *in situ* measurements is non-uniform across space, time and platforms. Different platforms employ a large variety of sensors, which operate in a wide range of often hostile environments and use different measurement protocols. Occasionally, buoys are left unattended for extended periods of time, while ships may involve a certain amount of the human-related impacts in data collection and transmission. Therefore, quality control is necessary before in situ data can be further used in scientific research or other applications. This is an example of quality control and monitoring of sea surface temperatures measured by ships and buoys, the iQuam system developed at NOAA/NESDIS/STAR, where statistics show the quality of *in situ* measurements of sea surface temperatures.

One of the problems facing real-time ocean observatories is the ability to provide a fast and accurate assessment of the data quality. Ocean Networks Canada is in the process of implementing real-time quality control on incoming data. For scalar data, the aim is to meet the guidelines of the Quality Assurance of Real Time Oceanographic Data (QARTOD) group. QARTOD is a US organization tasked with identifying issues involved with incoming real-time data from the U.S Integrated Ocean Observing System (IOOS). A large portion of their agenda is to create guidelines for how the quality of real-time data is to be determined and reported to the scientific community. Real-time data quality testing at Ocean Networks Canada includes tests designed to catch instrument failures and major spikes or data dropouts before the data is made available to the user. Real-time quality tests include meeting instrument manufacturer's standards and overall observatory/site ranges determined from previous data. Due to the positioning of some instrument platforms in highly productive areas, we have also designed dual-sensor tests e.g. for some conductivity sensors. The quality control testing is split into 3 separate categories. The first category is in real-time and tests the data before the data are parsed into the database. The second category is delayed-mode testing where archived data are subject to testing after a certain period of time. The third category is manual quality control by an Ocean Networks Canada data expert.

Historical Data Available

OceanSITES manages a set of links to various sources of available ocean data, including: the Hawaiian Ocean Timeseries (HOT), the JAMSTEC Kuroshio Extension Observatory (JKEO), Line W monitoring the North Atlantic's deep western boundary current, and others.

This site includes links to the ARGO Float Data, The Data Library and Archives (DLA), the Falmouth Monthly Climate Reports, Martha's Vineyard Coastal Observatory, the Multibeam Archive, the Seafloor Data and Observation Visualization Environment (SeaDOVE): A Web-served GIS Database of Multi-scalar Seafloor Data, Seafloor Sediments Data Collection, the Upper Ocean Mooring Data Archive, the U.S. GLOBEC Data System, U.S. JGOFS Data System, and the WHOI Ship Data-Grabber System.

There are a variety of data sets in a data library listed at Columbia University:

This library includes:

- LEVITUS94 is the World Ocean Atlas as of 1994, an atlas of objectively analyzed fields of major ocean parameters at the annual, seasonal, and monthly time scales. It is superseded by WOA98.

- NOAA NODC WOA98 is the World Ocean Atlas as of 1998, an atlas of objectively analyzed fields of major ocean parameters at monthly, seasonal, and annual time scales. Superseded by WOA01.

- NOAA NODC WOA01 is the World Ocean Atlas 2001, an atlas of objectively analyzed fields of major ocean parameters at monthly, seasonal, and annual time scales. Replaced by WOA05.

- NOAA NODC WOA05 is the World Ocean Atlas 2005, an atlas of objectively analyzed fields of major ocean parameters at monthly, seasonal, and annual time scales.

In situ observations spanning from the early 1700s to present are available from the International Comprehensive Ocean Atmosphere Data Set (ICOADS).

This data set includes observations of a number of the surface ocean and atmospheric variables from ships, moored and drifting buoys and C-MAN stations.

In 2006, Ocean Networks Canada began collecting high-resolution in-situ measurements from the seafloor in Saanich Inlet, near Victoria, British Columbia, Canada. Monitoring sites were later extended to the Strait of Georgia and 5 locations off the West coast of Vancouver Island, British Columbia, Canada. All historical measurements are freely available via Ocean Networks Canada's data portal, Oceans 2.0.

Ocean Reanalysis

Ocean reanalysis is a method of combining historical ocean observations with a general ocean model (typically a computational model) driven by historical estimates of surface winds, heat, and freshwater, by way of a data assimilation algorithm to reconstruct historical changes in the state of the ocean.

Historical observations are sparse and insufficient for understanding the history of the ocean and its circulation. By utilizing data assimilation techniques in combination with advanced computational models of the global ocean, researchers are able to interpolate the historical observations to all points in the ocean. This process has an analog in the construction of atmospheric reanalysis and is closely related to ocean state estimation.

Current Projects

A number of efforts have been initiated in recent years to apply data assimilation to estimate the physical state of the ocean, including temperature, salinity, currents, and sea level, in recent years. There are three alternative state estimation approaches. The first approach is used by the 'no-model' analyses, for which temperature or salinity observations update a first guess provided by climatological monthly estimates.

The second approach is that of the sequential data assimilation analyses, which move forward in time from a previous analysis using a numerical simulation of the evolving temperature and other variables produced by an ocean general circulation model. The simulation provides the first guess of the state of the ocean at the next analysis time, while corrections are made to this first guess based on observations of variables such as temperature, salinity, or sea level.

The third approach is 4D-Var, which in the implementation described uses the initial conditions and surface forcing as control variables to be modified in order to be consistent with the observations as well as a numerical representation of the equations of motion through iterative solution of a giant optimization problem.

Methodologies

No-model Approach

ISHII and LEVITUS begin with a first guess of the climatological monthly upper-ocean temperature based on climatologies produced by the NOAA National Oceanographic Data Center. The innovations are mapped onto the analysis levels. ISHII uses and alternative 3DVAR approach to do an objective mapping with a smaller decorrelation scale in midlatitudes (300 km) that elongates in the zonal direction by a factor of 3 at equatorial latitudes. LEVITUS begins similarly to ISHII, but uses the technique of Cressman and Barnes with a homogeneous scale of 555 km to objectively map the temperature innovation onto a uniform grid.

Sequential Approaches

The sequential approaches can be further divided into those using Optimal Interpolation and its more sophisticated cousin the Kalman Filter, and those using 3D-Var. Among those mentioned above, INGV and SODA use versions of Optimal Interpolation. CERFACS, GODAS, and GFDL all use 3DVar. "To date we are unaware of any attempt to use Kalman Filter for multi-decadal ocean reanalyses." The 4-Dimensional Local Ensemble Transform Kalman Filter (4D-LETKF) has been applied to the Geophysical Fluid Dynamics Laboratory's (GFDL) Modular Ocean Model (MOM2) for a 7-year ocean reanalysis from January 1997 – 2004.

Variational (4D-Var) Approach

One innovative attempt by GECCO has been made to apply 4D-Var to the decadal ocean estimation problem. This approach faces daunting computational challenges, but provides some interesting benefits including satisfying some conservation laws and the construction of the ocean model adjoint.

References

- Hunt, B.R., Kostelich E.J., Szunyogh, I. Efficient Data Assimilation for Spatiotemporal Chaos: A Local Ensemble Transform Kalman Filter. arXiv:physics/0511236 v1 28 Nov 2005. Dated 19, June 2020

- Jenkyns, Reyna (20 September 2010). "NEPTUNE Canada: Data integrity from the seafloor to your (Virtual) Door". Oceans 2010. doi:10.1109/OCEANS.2010.5664290. Retrieved 21, April 2020

- "The Physical Oceanography Component of Hawaii Ocean Timeseries (HOT/PO)". Soest.hawaii.edu. Retrieved 17, February 2020

Chapter 7

Effect of Pollution on Oceans

Pollution in oceans can be caused by humans when industrial, agricultural or residential waste enters the ocean and causes harm to the marine life and ecosystem. It can also be caused by noise, excessive amount of carbon dioxide, or the entry of invasive organisms. This chapter closely examines the effect of pollution on oceans to provide an extensive understanding of the subject.

Marine Pollution

Marine pollution occurs when harmful, or potentially harmful, effects result from the entry into the ocean of chemicals, particles, industrial, agricultural and residential waste, noise, or the spread of invasive organisms. Eighty percent of marine pollution comes from land. Air pollution is also a contributing factor by carrying off pesticides or dirt into the ocean. Land and air pollution have proven to be harmful to marine life and its habitats.

The pollution often comes from non point sources such as agricultural runoff, wind-blown debris and dust. Nutrient pollution, a form of water pollution, refers to contamination by excessive inputs of nutrients. It is a primary cause of eutrophication of surface waters, in which excess nutrients, usually nitrogen or phosphorus, stimulate algae growth.

Many potentially toxic chemicals adhere to tiny particles which are then taken up by plankton and benthos animals, most of which are either deposit or filter feeders. In this way, the toxins are concentrated upward within ocean food chains. Many particles combine chemically in a manner highly depletive of oxygen, causing estuaries to become anoxic.

When pesticides are incorporated into the marine ecosystem, they quickly become absorbed into marine food webs. Once in the food webs, these pesticides can cause mutations, as well as diseases, which can be harmful to humans as well as the entire food web.

Toxic metals can also be introduced into marine food webs. These can cause a change to tissue matter, biochemistry, behaviour, reproduction, and suppress growth in marine life. Also, many animal feeds have a high fish meal or fish hydrolysate content. In this way, marine toxins can be transferred to land animals, and appear later in meat and dairy products.

History

Although marine pollution has a long history, significant international laws to counter it were only enacted in the twentieth century. Marine pollution was a concern during several United Nations Conferences on the Law of the Sea beginning in the 1950s. Most scientists believed that the oceans were so vast that they had unlimited ability to dilute, and thus render pollution harmless.

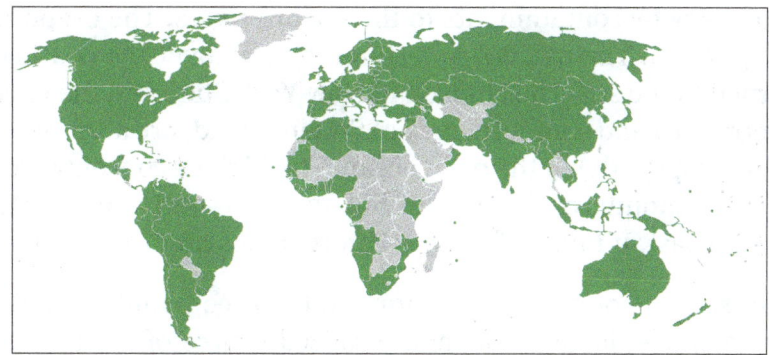

Parties to the MARPOL 73/78 convention on marine pollution.

In the late 1950s and early 1960s, there were several controversies about dumping radioactive waste off the coasts of the United States by companies licensed by the Atomic Energy Commission, into the Irish Sea from the British reprocessing facility at Windscale, and into the Mediterranean Sea by the French Commissariat à l'Energie Atomique. After the Mediterranean Sea controversy, for example, Jacques Cousteau became a worldwide figure in the campaign to stop marine pollution. Marine pollution made further international headlines after the 1967 crash of the oil tanker Torrey Canyon, and after the 1969 Santa Barbara oil spill off the coast of California.

Marine pollution was a major area of discussion during the 1972 United Nations Conference on the Human Environment, held in Stockholm. That year also saw the signing of the Convention on the Prevention of Marine Pollution by Dumping of Wastes and Other Matter, sometimes called the London Convention. The London Convention did not ban marine pollution, but it established black and gray lists for substances to be banned (black) or regulated by national authorities (gray). Cyanide and high-level radioactive waste, for example, were put on the black list. The London Convention applied only to waste dumped from ships, and thus did nothing to regulate waste discharged as liquids from pipelines.

Pathways of Pollution

Septic river.

There are many different ways to categorize, and examine the inputs of pollution into our marine ecosystems. Patin (n.d.) notes that generally there are three main types of inputs of pollution into the ocean: direct discharge of waste into the oceans, runoff into the waters due to rain, and pollutants that are released from the atmosphere.

One common path of entry by contaminants to the sea are rivers. The evaporation of water from oceans exceeds precipitation. The balance is restored by rain over the continents entering rivers and then being returned to the sea. The Hudson in New York State and the Raritan in New Jersey, which empty at the northern and southern ends of Staten Island, are a source of mercury contamination of zooplankton (copepods) in the open ocean. The highest concentration in the filter-feeding copepods is not at the mouths of these rivers but 70 miles south, nearer Atlantic City, because water flows close to the coast. It takes a few days before toxins are taken up by the plankton.

Pollution is often classed as point source or nonpoint source pollution. Point source pollution occurs when there is a single, identifiable, and localized source of the pollution. An example is directly discharging sewage and industrial waste into the ocean. Pollution such as this occurs particularly in developing nations. Nonpoint source pollution occurs when the pollution comes from ill-defined and diffuse sources. These can be difficult to regulate. Agricultural runoff and wind blown debris are prime examples.

Direct Discharge

Acid mine drainage in the Rio Tinto River.

Pollutants enter rivers and the sea directly from urban sewerage and industrial waste discharges, sometimes in the form of hazardous and toxic wastes.

Inland mining for copper, gold. etc., is another source of marine pollution. Most of the pollution is simply soil, which ends up in rivers flowing to the sea. However, some minerals discharged in the course of the mining can cause problems, such as copper, a common industrial pollutant, which can interfere with the life history and development of coral polyps. Mining has a poor environmental track record. For example, according to the United States Environmental Protection Agency, mining has contaminated portions of the headwaters of over 40% of watersheds in the western continental US. Much of this pollution finishes up in the sea.

Land Runoff

Surface runoff from farming, as well as urban runoff and runoff from the construction of roads, buildings, ports, channels, and harbours, can carry soil and particles laden with carbon, nitrogen, phosphorus, and minerals. This nutrient-rich water can cause fleshy algae and phytoplankton to thrive in coastal areas; known as algal blooms, which have the potential to create hypoxic conditions by using all available oxygen.

Polluted runoff from roads and highways can be a significant source of water pollution in coastal areas. About 75% of the toxic chemicals that flow into Puget Sound are carried by stormwater that runs off paved roads and driveways, rooftops, yards and other developed land.

Ship Pollution

A cargo ship pumps ballast water over the side.

Ships can pollute waterways and oceans in many ways. Oil spills can have devastating effects. While being toxic to marine life, polycyclic aromatic hydrocarbons (PAHs), found in crude oil, are very difficult to clean up, and last for years in the sediment and marine environment.

Oil spills are probably the most emotive of marine pollution events. However, while a tanker wreck may result in extensive newspaper headlines, much of the oil in the world's seas comes from other smaller sources, such as tankers discharging ballast water from oil tanks used on return ships, leaking pipelines or engine oil disposed of down sewers.

Discharge of cargo residues from bulk carriers can pollute ports, waterways and oceans. In many instances vessels intentionally discharge illegal wastes despite foreign and domestic regulation prohibiting such actions. It has been estimated that container ships lose over 10,000 containers at sea each year (usually during storms). Ships also create noise pollution that disturbs natural wildlife, and water from ballast tanks can spread harmful algae and other invasive species.

Ballast water taken up at sea and released in port is a major source of unwanted exotic marine life. The invasive freshwater zebra mussels, native to the Black, Caspian and Azov seas, were probably transported to the Great Lakes via ballast water from a transoceanic vessel. Meinesz believes that one of the worst cases of a single invasive species causing harm to an ecosystem can be attributed to a seemingly harmless jellyfish. *Mnemiopsis leidyi*, a species of comb jellyfish that spread so it now inhabits estuaries in many parts of the world. It was first introduced in 1982, and thought to have been transported to the Black Sea in a ship's ballast water. The population of the jellyfish shot up exponentially and, by 1988, it was wreaking havoc upon the local fishing industry. "The anchovy catch fell from 204,000 tons in 1984 to 200 tons in 1993; sprat from 24,600 tons in 1984 to 12,000 tons in 1993; horse mackerel from 4,000 tons in 1984 to zero in 1993." Now that the jellyfish have exhausted the zooplankton, including fish larvae, their numbers have fallen dramatically, yet they continue to maintain a stranglehold on the ecosystem.

Invasive species can take over once occupied areas, facilitate the spread of new diseases, introduce new genetic material, alter underwater seascapes and jeopardize the ability of native species to obtain food. Invasive species are responsible for about $138 billion annually in lost revenue and management costs in the US alone.

Atmospheric Pollution

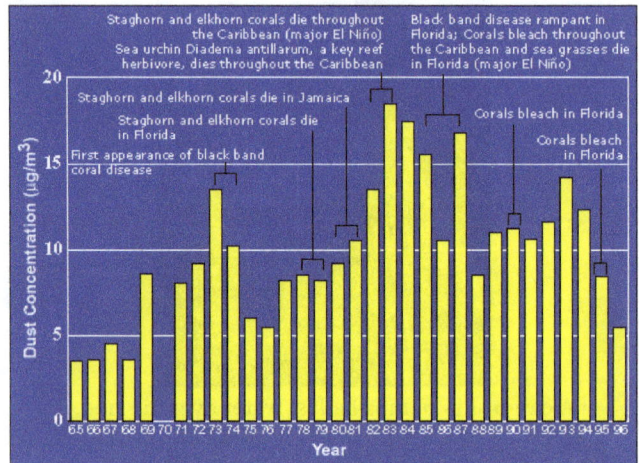

Graph linking atmospheric dust to various coral deaths across the Caribbean Sea and Florida.

Another pathway of pollution occurs through the atmosphere. Wind blown dust and debris, including plastic bags, are blown seaward from landfills and other areas. Dust from the Sahara moving around the southern periphery of the subtropical ridge moves into the Caribbean and Florida during the warm season as the ridge builds and moves northward through the subtropical Atlantic. Dust can also be attributed to a global transport from the Gobi and Taklamakan deserts across Korea, Japan, and the Northern Pacific to the Hawaiian Islands. Since 1970, dust outbreaks have worsened due to periods of drought in Africa. There is a large variability in dust transport to the Caribbean and Florida from year to year; however, the flux is greater during positive phases of the North Atlantic Oscillation. The USGS links dust events to a decline in the health of coral reefs across the Caribbean and Florida, primarily since the 1970s.

Climate change is raising ocean temperatures and raising levels of carbon dioxide in the atmosphere. These rising levels of carbon dioxide are acidifying the oceans. This, in turn, is altering aquatic ecosystems and modifying fish distributions, with impacts on the sustainability of fisheries and the livelihoods of the communities that depend on them. Healthy ocean ecosystems are also important for the mitigation of climate change.

Deep Sea Mining

Deep sea mining is a relatively new mineral retrieval process that takes place on the ocean floor. Ocean mining sites are usually around large areas of polymetallic nodules or active and extinct hydrothermal vents at about 1,400 – 3,700 meters below the ocean's surface. The vents create sulfide deposits, which contain precious metals such as silver, gold, copper, manganese, cobalt, and zinc. The deposits are mined using either hydraulic pumps or bucket systems that take ore to the surface to be processed. As with all mining operations, deep sea mining raises questions about environmental damages to the surrounding areas

Because deep sea mining is a relatively new field, the complete consequences of full scale mining operations are unknown. However, experts are certain that removal of parts of the sea floor will result in disturbances to the benthic layer, increased toxicity of the water column and sediment plumes from tailings. Removing parts of the sea floor disturbs the habitat of benthic organisms, possibly, depending on the type of mining and location, causing permanent disturbances. Aside from direct impact of mining the area, leakage, spills and corrosion would alter the mining area's chemical makeup.

Among the impacts of deep sea mining, sediment plumes could have the greatest impact. Plumes are caused when the tailings from mining (usually fine particles) are dumped back into the ocean, creating a cloud of particles floating in the water. Two types of plumes occur: near bottom plumes and surface plumes. Near bottom plumes occur when the tailings are pumped back down to the mining site. The floating particles increase the turbidity, or cloudiness, of the water, clogging filter-feeding apparatuses used by benthic organisms. Surface plumes cause a more serious problem. Depending on the size of the particles and water currents the plumes could spread over vast areas. The plumes could impact zooplankton and light penetration, in turn affecting the food web of the area.

Types of Pollution

Acidification

Island with fringing reef in the Maldives. Coral reefs are dying around the world.

The oceans are normally a natural carbon sink, absorbing carbon dioxide from the atmosphere. Because the levels of atmospheric carbon dioxide are increasing, the oceans are becoming more acidic. The potential consequences of ocean acidification are not fully understood, but there are concerns that structures made of calcium carbonate may become vulnerable to dissolution, affecting corals and the ability of shellfish to form shells.

Oceans and coastal ecosystems play an important role in the global carbon cycle and have removed about 25% of the carbon dioxide emitted by human activities between 2000 and 2007 and about half the anthropogenic CO_2 released since the start of the industrial revolution. Rising ocean temperatures and ocean acidification means that the capacity of the ocean carbon sink will gradually get weaker, giving rise to global concerns expressed in the Monaco and Manado Declarations.

A report from NOAA scientists published in the journal Science in May 2008 found that large amounts of relatively acidified water are upwelling to within four miles of the Pacific continental shelf area of North America. This area is a critical zone where most local marine life lives or is born. While the paper dealt only with areas from Vancouver to northern California, other continental shelf areas may be experiencing similar effects.

A related issue is the methane clathrate reservoirs found under sediments on the ocean floors. These trap large amounts of the greenhouse gas methane, which ocean warming has the potential to release. In 2004 the global inventory of ocean methane clathrates was estimated to occupy between one and five million cubic kilometres. If all these clathrates were to be spread uniformly across the ocean floor, this would translate to a thickness between three and fourteen metres. This estimate corresponds to 500–2500 gigatonnes carbon (Gt C), and can be compared with the 5000 Gt C estimated for all other fossil fuel reserves.

Eutrophication

Polluted lagoon.

Effect of eutrophication on marine benthic life.

Eutrophication is an increase in chemical nutrients, typically compounds containing nitrogen or phosphorus, in an ecosystem. It can result in an increase in the ecosystem's primary productivity (excessive plant growth and decay), and further effects including lack of oxygen and severe reductions in water quality, fish, and other animal populations.

The biggest culprit are rivers that empty into the ocean, and with it the many chemicals used as fertilizers in agriculture as well as waste from livestock and humans. An excess of oxygen depleting chemicals in the water can lead to hypoxia and the creation of a dead zone.

Estuaries tend to be naturally eutrophic because land-derived nutrients are concentrated where runoff enters the marine environment in a confined channel. The World Resources Institute has identified 375 hypoxic coastal zones around the world, concentrated in coastal areas in Western Europe, the Eastern and Southern coasts of the US, and East Asia, particularly in Japan. In the ocean, there are frequent red tide algae blooms that kill fish and marine mammals and cause respiratory problems in humans and some domestic animals when the blooms reach close to shore.

In addition to land runoff, atmospheric anthropogenic fixed nitrogen can enter the open ocean. A study in 2008 found that this could account for around one third of the ocean's external (non-recycled) nitrogen supply and up to three per cent of the annual new marine biological production. It has been suggested that accumulating reactive nitrogen in the environment may have consequences as serious as putting carbon dioxide in the atmosphere.

One proposed solution to eutrophication in estuaries is to restore shellfish populations, such as oysters. Oyster reefs remove nitrogen from the water column and filter out suspended solids, subsequently reducing the likelihood or extent of harmful algal blooms or anoxic conditions. Filter feeding activity is considered beneficial to water quality by controlling phytoplankton density and sequestering nutrients, which can be removed from the system through shellfish harvest, buried in the sediments, or lost through denitrification. Foundational work toward the idea of improving marine water quality through shellfish cultivation to was conducted by Odd Lindahl et al., using mussels in Sweden.

Plastic Debris

A mute swan builds a nest using plastic garbage.

Marine debris is mainly discarded human rubbish which floats on, or is suspended in the ocean. Eighty percent of marine debris is plastic – a component that has been rapidly accumulating since the end of World War II. The mass of plastic in the oceans may be as high as 100,000,000 tonnes (98,000,000 long tons; 110,000,000 short tons).

Discarded plastic bags, six pack rings and other forms of plastic waste which finish up in the ocean present dangers to wildlife and fisheries. Aquatic life can be threatened through entanglement, suffocation, and ingestion. Fishing nets, usually made of plastic, can be left or lost in the ocean by fishermen. Known as ghost nets, these entangle fish, dolphins, sea turtles, sharks, dugongs, crocodiles, seabirds, crabs, and other creatures, restricting movement, causing starvation, laceration and infection, and, in those that need to return to the surface to breathe, suffocation.

Remains of an albatross containing ingested flotsam.

Many animals that live on or in the sea consume flotsam by mistake, as it often looks similar to their natural prey. Plastic debris, when bulky or tangled, is difficult to pass, and may become permanently lodged in the digestive tracts of these animals. Especially when evolutionary adaptions make it impossible for the likes of turtles to reject plastic bags, which resemble jellyfish when immersed in water, as they have a system in their throat to stop slippery foods from otherwise escaping. Thereby blocking the passage of food and causing death through starvation or infection.

Plastics accumulate because they don't biodegrade in the way many other substances do. They will photodegrade on exposure to the sun, but they do so properly only under dry conditions, and water inhibits this process. In marine environments, photodegraded plastic disintegrates into ever smaller pieces while remaining polymers, even down to the molecular level. When floating plastic particles photodegrade down to zooplankton sizes, jellyfish attempt to consume them, and in this way the plastic enters the ocean food chain. Many of these long-lasting pieces end up in the stomachs of marine birds and animals, including sea turtles, and black-footed albatross.

Plastic debris tends to accumulate at the centre of ocean gyres. In particular, the Great Pacific Garbage Patch has a very high level of plastic particulate suspended in the upper water column. In samples taken in 1999, the mass of plastic exceeded that of zooplankton (the dominant animal life in the area) by a factor of six. Midway Atoll, in common with all the Hawaiian Islands, receives substantial amounts of debris from the garbage patch. Ninety percent plastic, this debris accumulates on the beaches of Midway where it becomes a hazard to the bird population of the island. Midway Atoll is home to two-thirds (1.5 million) of the global population of Laysan albatross. Nearly all of these albatross have plastic in their digestive system and one-third of their chicks die.

Marine debris on Kamilo Beach, Hawaii, washed up from the Great Pacific Garbage Patch.

Toxic additives used in the manufacture of plastic materials can leach out into their surroundings when exposed to water. Waterborne hydrophobic pollutants collect and magnify on the surface of plastic debris, thus making plastic far more deadly in the ocean than it would be on land. Hydrophobic contaminants are also known to bioaccumulate in fatty tissues, biomagnifying up the food chain and putting pressure on apex predators. Some plastic additives are known to disrupt the endocrine system when consumed, others can suppress the immune system or decrease reproductive rates. Floating debris can also absorb persistent organic pollutants from seawater, including PCBs, DDT and PAHs. Aside from toxic effects, when ingested some of these are mistaken by the animal brain for estradiol, causing hormone disruption in the affected wildlife.

Toxins

Apart from plastics, there are particular problems with other toxins that do not disintegrate rapidly in the marine environment. Examples of persistent toxins are PCBs, DDT, TBT, pesticides, furans, dioxins, phenols and radioactive waste. Heavy metals are metallic chemical elements that have a relatively high density and are toxic or poisonous at low concentrations. Examples are mercury, lead, nickel, arsenic and cadmium. Such toxins can accumulate in the tissues of many species of aquatic life in a process called bioaccumulation. They are also known to accumulate in benthic environments, such as estuaries and bay muds: a geological record of human activities of the last century.

Specific Examples

- Chinese and Russian industrial pollution such as phenols and heavy metals in the Amur River have devastated fish stocks and damaged its estuary soil.

- Wabamun Lake in Alberta, Canada, once the best whitefish lake in the area, now has unacceptable levels of heavy metals in its sediment and fish.

- Acute and chronic pollution events have been shown to impact southern California kelp forests, though the intensity of the impact seems to depend on both the nature of the contaminants and duration of exposure.

- Due to their high position in the food chain and the subsequent accumulation of heavy metals from their diet, mercury levels can be high in larger species such as bluefin and albacore. As a result, in March 2004 the United States FDA issued guidelines recommending that pregnant women, nursing mothers and children limit their intake of tuna and other types of predatory fish.

- Some shellfish and crabs can survive polluted environments, accumulating heavy metals or toxins in their tissues. For example, mitten crabs have a remarkable ability to survive in highly modified aquatic habitats, including polluted waters. The farming and harvesting of such species needs careful management if they are to be used as a food.

- Surface runoff of pesticides can alter the gender of fish species genetically, transforming male into female fish.

- Heavy metals enter the environment through oil spills – such as the Prestige oil spill on the Galician coast – or from other natural or anthropogenic sources.

- In 2005, the 'Ndrangheta, an Italian mafia syndicate, was accused of sinking at least 30 ships loaded with toxic waste, much of it radioactive. This has led to widespread investigations into radioactive-waste disposal rackets.

- Since the end of World War II, various nations, including the Soviet Union, the United Kingdom, the United States, and Germany, have disposed of chemical weapons in the Baltic Sea, raising concerns of environmental contamination.

Underwater Noise

Marine life can be susceptible to noise or the sound pollution from sources such as passing ships, oil exploration seismic surveys, and naval low-frequency active sonar. Sound travels more rapidly and over larger distances in the sea than in the atmosphere. Marine animals, such as cetaceans, often have weak eyesight, and live in a world largely defined by acoustic information. This applies also to many deeper sea fish, who live in a world of darkness. Between 1950 and 1975, ambient noise at one location in the Pacific Ocean increased by about ten decibels (that is a tenfold increase in intensity).

Noise also makes species communicate louder, which is called the Lombard vocal response. Whale songs are longer when submarine-detectors are on. If creatures don't "speak" loud enough, their voice can be masked by anthropogenic sounds. These unheard voices might be warnings, finding of prey, or preparations of net-bubbling. When one species begins speaking louder, it will mask other species voices, causing the whole ecosystem to eventually speak louder.

According to the oceanographer Sylvia Earle, "Undersea noise pollution is like the death of a thousand cuts. Each sound in itself may not be a matter of critical concern, but taken all together, the noise from shipping, seismic surveys, and military activity is creating a totally different environment than existed even 50 years ago. That high level of noise is bound to have a hard, sweeping impact on life in the sea."

Adaptation and Mitigation

Aerosol can polluting a beach.

Much anthropogenic pollution ends up in the ocean. The 2011 edition of the United Nations Environment Programme Year Book identifies as the main emerging environmental issues the loss to the oceans of massive amounts of phosphorus, "a valuable fertilizer needed to feed a growing global population", and the impact billions of pieces of plastic waste are having globally on the health of marine environments. Bjorn Jennssen (2003) notes in his article, "Anthropogenic pollution may reduce biodiversity and productivity of marine ecosystems, resulting in reduction and depletion of human marine food resources". There are two ways the overall level of this pollution can be mitigated: either the human population is reduced, or a way is found to reduce the ecological footprint left behind by the average human. If the second way is not adopted, then the first way may be imposed as world ecosystems falter.

The second way is for humans, individually, to pollute less. That requires social and political will, together with a shift in awareness so more people respect the environment and are less disposed to abuse it. At an operational level, regulations, and international government participation is needed. It is often very difficult to regulate marine pollution because pollution spreads over international barriers, thus making regulations hard to create as well as enforce.

Without appropriate awareness of marine pollution, the necessary global will to effectively address the issues may prove inadequate. Balanced information on the sources and harmful effects of marine pollution need to become part of general public awareness, and ongoing research is required to fully establish, and keep current, the scope of the issues. As expressed in Daoji and Dag's research, one of the reasons why environmental concern is lacking among the Chinese is because the public awareness is low and therefore should be targeted. Likewise, regulation, based upon such in-depth research should be employed. In California, such regulations have already been put in place to protect Californian coastal waters from agricultural runoff. This includes the California Water Code, as well as several voluntary programs. Similarly, in India, several tactics have been employed that help reduce marine pollution, however, they do not significantly target the problem.

In Chennai, sewage has been dumped further into open waters. Due to the mass of waste being deposited, open-ocean is best for diluting, and dispersing pollutants, thus making them less harmful to marine ecosystems.

References

- Warner R (2009) Protecting the oceans beyond national jurisdiction: strengthening the international law framework. Vol. 3 of Legal aspects of sustainable development, Brill, ISBN 978-90-04-17262-3

- "Midway's albatross population stable | Hawaii's Newspaper". The Honolulu Advertiser. 17 January 2005. Retrieved 20 May 2012

- Moore, Charles (November 2003). "Across the Pacific Ocean, plastics, plastics, everywhere". Natural History. Archived from the original on 27 September 2007. Retrieved 16, January 2020

- "Research | AMRF/ORV Alguita Research Projects" Algalita Marine Research Foundation. Macdonald Design. Retrieved 01, June 2020

- Weiss, Kenneth R. (2 August 2006). "Plague of Plastic Chokes the Seas". Los Angeles Times. Archived from the original on 25 March 2008. Retrieved 19, August 2020

- Hamblin, Jacob Darwin (2008) Poison in the Well: Radioactive Waste in the Oceans at the Dawn of the Nuclear Age. Rutgers University Press. ISBN 978-0-8135-4220-1

- Administration, US Department of Commerce, National Oceanic and Atmospheric. "What is the biggest source of pollution in the ocean?". oceanservice.noaa.gov. Retrieved 18, February 2020

- Podsadam, Janice (19 June 2001). "Lost Sea Cargo: Beach Bounty or Junk?". National Geographic News. Retrieved 19, April 2020

Permissions

Index

CPSIA information can be obtained
at www.ICGtesting.com
Printed in the USA
BVHW012039300822
645853BV00002B/87